RADIATION SAFETY PROCEDURES AND TRAINING FOR THE RADIATION SAFETY OFFICER

ALSO BY JOHN R HAYGOOD

The Terrorist Effect: Weapons of Mass Disruption—
The Danger of Nuclear Terrorism
by James William Jones and John R Haygood
ISBN: 978-1-4620-3932-6

RADIATION SAFETY PROCEDURES AND TRAINING FOR THE RADIATION SAFETY OFFICER

GUIDANCE FOR PREPARING A RADIATION SAFETY PROGRAM

JOHN R HAYGOOD

iUniverse LLC
Bloomington

Radiation Safety Procedures and Training for the Radiation Safety Officer
Guidance for Preparing a Radiation Safety Program

iUniverse books may be ordered through booksellers or by contacting:

iUniverse LLC
1663 Liberty Drive
Bloomington, IN 47403
www.iuniverse.com
1-800-Authors (1-800-288-4677)

ISBN: 978-1-4917-0596-4 (sc)
ISBN: 978-1-4917-0597-1 (ebk)

Printed in the United States of America

iUniverse rev. date: 09/11/2013

Contents

Illustrations

Tables

Preface

This book is designed to provide radiation safety officers and users/operators of devices using radiation with the tools needed to operate a safe program, construct training materials and courses, AND to comply with regulatory requirements. It is centered primarily around radioactive materials license requirements, but much of the material can be applied to non-healing art x-ray, accelerator, and laser operations and registrations. All of the information consists of either original text created by the author or compilations of regulatory information/requirements and of common knowledge scientific information found in standard tables and references. Comments, suggestions, and recommendations from the reader will be appreciated. This is not a health physics production. Only a minimal amount of radiation principles are provided to provide the reader/user with enough information to proceed through the material. *The author recommends that the serious user, desiring a greater degree of knowledge in health physics and radiation principles, obtain and study the references listed in this introductory section.*

While the procedures presented stem predominantly from the Texas radiation regulatory program, the book should be applicable in any program operated in any state, except that references to specific regulations will be different. All states are either regulated by the US Nuclear Regulatory Commission (NRC) or have an "Agreement State" status (Agreement State is explained later in the book). In any case, the laws and rules are approximately the same. Under NRC requirements, all agreement states must have rules "compatible" with those of the NRC. This is discussed in more detail in the text.

CAUTION: References to and quotes from the Texas Regulations for Control of Radiation (TRCR) and from 25 Texas Administrative Code 289 are subject to change and to interpretation by the (Texas) Department of State Health Services (DSHS). Interpretations of the TRCR are the purview of the TDH and the Texas courts. In some regulatory areas, the Texas Commission on Environmental Quality or the Texas Railroad Commission (TRRC) have jurisdiction. Further, references to regulations and procedures of the US DOT and the US NRC are also subject to change and interpretation by those agencies and federal courts. The concepts and information provided here are the opinions and concepts of the author. The user must adhere to the official interpretations and guidance of the agency with jurisdiction (Agreement State or federal government).

About the author

John Haygood retired in 1997 from the Texas radiation control program managed by the Bureau of Radiation Control (BRC) of the Texas Department of Health (now DSHS) following 25 years of public service. He also worked for the program from 2002 to 2008. During his time of service, he performed many inspections of radioactive materials licensed operations, including Increased Controls security inspections, and registered x-ray and laser use operations, and investigated numerous radiation incidents. He was also extensively involved in the development of many of the current inspection, compliance, enforcement, licensing and registration processes—as well as developing inspector training programs and procedures. Additionally, he has dealt extensively with legal, emergency response, NORM, uranium license, transportation, low level waste, and security processes.

Mr. Haygood would like to continue serving the radiation using public through sharing his experience to help maintain a safe radiation environment.

Acknowledgements

The author is grateful to the following companies and individuals who graciously reviewed the material and provided comments and/or contributed information incorporated into the manual:

Companies

Ludlum Measurements, Inc.

501 Oak Street
PO Box 810
Sweetwater, Texas 79556
Phone: 800-622-0828 (USA)
Fax: 325-235-4672
Email: ludlum@ludlums.com
Website: http://www.ludlums.com

Health Physicists

Charles R. Meyer, CHP
Round Rock, Texas

Eric Skotak
Austin, Texas

Security

James Wm. Jones Ph.D., P.E
Huntington Beach, California

Thank you to my wife, Jesusa Haygood, for her support in putting this document together.

Chapter 1

Introduction to Radiation

1. Introduction

This book is intended to provide an understanding of radiation principles and radiation safety practices at a level which will help "lay persons" protect themselves and others from unnecessary radiation exposure during operations which require the use of radiation sources. It is not intended to develop radiation "experts" or health physicists as that would require considerably more material and study. Nor is it intended to train persons in the operation of specific equipment. Consequently, it has been prepared with the assumption that persons studying this material have at least the equivalent of a high school education and that math and science may not have been their "strong points". With a few "minor" exceptions, the math and science learned in high school should provide a sufficient background for the reader.

The book is also intended to provide the trainee/reader with the "tools" necessary to comply with the requirements of the radiation control regulatory authority or agency with jurisdiction. In Texas, the Department of State Health Services (DSHS) is the agency charged with enforcing the Texas Radiation Control Act—which is the radiation control enabling law in the state. However, the Texas Commission on Environmental Quality (TCEQ) has authority relating to waste disposal processes and the Texas Railroad Commission (TRRC) has authority related to oil field radioactive waste disposal. Since the rules/regulations in effect in most states are generally similar, if not identical, the procedures and practices should be applicable for the trainee/reader in any state. Generally, radiation control regulation is set within the state's health or environmental departments. Often, there is a combination of the two types of agencies with split duties and authorities.

The U. S. Nuclear Regulatory Commission (NRC) is the federal agency that regulates nuclear reactors and most of the uses of radioactive material in the United States.

Thirty-seven states have signed an agreement with the NRC allowing them to regulate the use of radioactive materials—except for nuclear reactors and federal operations. Several other states have applied for agreement status. The current agreement states, as of January 2012, are listed in Table 1-1. The NRC requires that each state implement "compatible" laws and regulations as part of the agreement. This process has developed a nationwide "regulatory system" which allows the use of radiation sources under similar practices throughout the country. The NRC recently began regulation of certain naturally occurring and accelerator produced radioactive materials, but does not regulate machine produced radiation. The U.S. Federal Drug Administration (FDA) regulates the latter at the manufacturing/distribution level. There are other organizations which are involved in the radiation regulatory process, and these will be discussed in Chapter 5.

State	Date of Agreement	State	Date of Agreement
Alabama	10/01/66	New Hampshire	05/16/65
Arizona	05/15/67	New Jersey	09/30/09
Arkansas	07/01/63	New Mexico	05/01/74
California	09/01/62	New York	10/15/62
Colorado	02/01/68	North Carolina	08/01/64
Florida	07/01/64	North Dakota	09/01/69
Georgia	12/15/69	Ohio	08/31/99
Illinois	06/01/87	Oklahoma	09/29/00
Iowa	01/01/86	Oregon	07/01/65
Kansas	01/01/65	Pennsylvania	03/31/08
Kentucky	03/26/62	Rhode Island	01/01/80
Louisiana	05/01/67	South Carolina	09/15/69
Maine	04/01/92	Tennessee	09/01/65
Maryland	01/01/71	Texas	03/01/63
Massachusetts	03/21/97	Utah	04/01/84
Minnesota	03/31/06	Virginia	03/31/09
Mississippi	07/01/62	Washington	12/31/66
Nebraska	10/01/66	Wisconsin	08/11/03
Nevada	07/01/72		

Table 1-1 List of Agreement States[1]

1 Conference of Radiation Control Program Directors, Inc. : http://www.crcpd.org/Map/ListAgreementStates (December 5, 2012)

The (potential) user of radiation sources should find this book useful in attaining a "comfort level" which will help him/her use radiation sources in a safe manner. The book should also be useful to the radiation safety officer (RSO) in most radiation use programs.

Basic Training Tools

In preparing for the use of radiation sources, each agency generally requires appropriate training be provided to the radiation safety officer and the radiation source users (radiation workers). This book is designed to help meet those requirements. Training can be provided by a training provider (in some cases the training provider must be approved) or the licensee/registrant. When developing one's own training course, in addition to this book, one should consider other training tools/resources, such as a workbook which addresses the areas applicable to the type of radiation being used and audio/visual aids (overhead projections, video tapes, movies, pictures, drawings, and mock-ups). The use of actual radiation sources/equipment is desirable, but not always necessary. After learning the "basics" of radiation safety, a good on-the-job-training program supervised by well-trained and experienced persons should properly equip the new radiation worker to operate safely.

2. Math Tools and Units

Scientific Notation—Brief Review

In many situations, dealing with radiation matters requires the use of very large and very small numbers and an extensive use of mathematics. Scientific notation allows us to handle large and small numbers more easily—especially when manipulating or converting units. In keeping with the intent of this book, a "low math" approach, scientific notation will be briefly reviewed here. Readers already comfortable with this math and the units can skip this section.

Our system of numbers is "base 10", which relies on multiples of "10". The number 10 is obviously one 10. It is also 1 times 10. The number 100 is ten 10's—or 10 times 10. This can be shown as:

$$\mathbf{10 \times 10\ (= 100)}$$

or

$$\mathbf{10^2\ (= 10 \times 10 = 100)}$$

The superscripted number "2" of the last 10 is termed an exponent.

Exponential Form

The number is represented in the exponential form. This form is often called scientific notation. Since $10^2 = 100$, and there are two zeroes in 100, then for each number which is a multiple of tens, an exponent can be used to express the 10 with the exponent showing the number of zeroes, such as:

100	=	10^2	(two zeroes in one-hundred)
1000	=	10^3	(three zeroes in one-thousand)
10,000	=	10^4	(four zeroes in ten-thousand)
100,000	=	10^5	(five zeroes in one-hundred-thousand)

and so on. Note that $10 = 10^1$ (one zero, the exponent is 1) and, further, $10^0 = 1$.

On the other hand, if a number is less than 1, the exponent will be negative. For example, $0.1 = 1/10$, which is $1 / 10^1$. By convention, this can be expressed as 10^{-1}. Further, 0.01 is 10^{-2}, 0.001 is 10^{-3}, and so on.

If 100 is multiplied by 1000, a result of 100,000 is yielded (it is, after all, one-hundred one-thousands). The equivalent exponential form can be substituted:

$$100 \text{ X } 1000 = 100{,}000$$

or

$$10^2 \text{ X } 10^3 = 10^5$$

Note that the exponent 2 of the first term plus the exponent 3 of the second term, added together result in the number 5 which is the exponent of the third term. This is because the product of a number expressed as an exponential number with exponent *a*, and a number expressed as an exponential number with exponent *b*, can be expressed as an exponential number with exponent *(a + b)*.

$$X^a \bullet X^b = Z^{(a+b)}$$

(using the convention that A • A represents multiplication)

Without further explanation, it should be noted for division the following is true:

$$X^a / X^b = Z^{(a-b)}$$

In other words, the exponent of the divisor can be subtracted from the exponent of the dividend when using scientific notation or exponential form. For example:

$$\mathbf{1000 / 100 = 10^3 / 10^2 = 10^{(3-2)} = 10^1 = 10}$$

To work with a number that is not exactly a multiple of ten requires a simple manipulation. For example, the number 1200.0 can be expressed as 1.2 X 1000. This is the same as 1.2 X 10^3. Note that if one counts from the position of the decimal to the left and repositions the decimal just to the right of the last figure (after the one but in front of the two), the result is 3. This is the same as the exponent of 10^3. Now if the number were smaller than 1, say 0.034, one would count the other way (to the right but to the same position) and show a NEGATIVE exponent.

$$\mathbf{0.034 = 3.4 \text{ X } 10^{-2}} \quad \text{(exponent of -2, counting from decimal to right)}$$

These steps can be put together to have:

$$\begin{aligned} \mathbf{1200.0 \text{ X } 0.034} &= \mathbf{1.2 \times 10^2 \text{ X } 3.4\ 10^{-2}} \\ &= \mathbf{1.2 \text{ X } 3.4 \text{ X } (10^3 \times 10^{-2})} \\ &= \mathbf{4.08 \text{ X } 10^1 = 40.8} \end{aligned}$$

Use of exponential form will allow one to work less painfully with very large and very small numbers. Radiation related units and terms rely heavily on this type of notation.

Significant Figures

Significant figures are the non-zero numbers to the right of the decimal. For 0.3480, the significant figures would be 3, 4, and 8.

Units and Terms

Although radiation units and terms will be discussed in a later chapter, certain basic units and terms should be reviewed (or learned) to assure better understanding.

Unit Modifiers

The difference between a meter and a kilometer is quite straightforward. Simply, one kilometer equals 1000 meters. The "kilo" prefix stands for "1000"—representing the multiples of tens. Using the prefix for the unit to show magnitude greatly simplifies communications and facilitates better understanding when dealing with science and math. Fortunately, there are a number of prefixes, taken from the Greek, which are universally accepted as unit modifiers.

Multiple of 10	Prefix	Symbol
10^{12}	tera	T
10^{9}	giga	G
10^{6}	mega	M
10^{3}	kilo	k
10^{1}		
10^{-1}	deci	d
10^{-2}	centi	c
10^{-3}	milli	m
10^{-6}	micro	F
10^{-9}	nano	n
10^{-12}	pico	p
10^{-15}	femto	f

Table 1-2 - Units and Subunits

Many of the prefixes above will be used frequently in radiation related matters.

Note: The author prefers to label units with prefixes as "subunits". For example the *kilometer* and *millimeter* would be **subunits** of the basic unit of **meter**.

Conversion of Common Units

Most persons educated in the United States learned the British system of measurement (foot, pound, second) as the primary system and metric as secondary. This is slowly being reversed. In matters of radiation, the SI (International System of Units) and MKS (metre – kilogram – second) systems are used more extensively—but there is a mixture of all systems in common use. Below is a table listing some of the standard or basic units of measurement and their conversion equivalent that may be used in this book.

Unit (Symbol)	Equivalent (Symbol)	Conversion from/to
meter (m)	39.37 inches (in)	SI/British
foot (ft)	.3048 m	British/SI
mile (mi)	1.6094 kilometers	British/SI
kilometer (km)	0.62137 mi	SI/British
kilogram (kg)	2.205 pounds	SI/British
pound (lb)	0.4536 kg	British/SI

Table 1-3 - Basic Unit Conversions

The following are the standard relationships between the Celsius and British systems.

Temperature Conversion (Note: C = Centigrade and F = Fahrenheit, Degrees C = EC and Degrees F = EF.)

Centigrade to Fahrenheit: **EC = (EF - 32) / 1.8**

Fahrenheit to Centigrade: **EF = 1.8 EC + 32**

3. Uses of Radiation

The medical, educational, and industrial uses of radiation are numerous and new techniques constantly appear. X-ray machines, radioactive materials, and lasers are used in so many ways that volumes would be required to describe them. Without these tools, our lives would be shorter and less healthful and our "quality of life" would be less than we currently enjoy. Surgical removal of a cancerous limb instead of radiation treatment, for example, would make a persons daily routine far more arduous. Many physical systems and objects that could be dangerous are made far safer by x-raying them to eliminate hidden faults. A brief review of the various uses follows.

Medical Uses

Diagnosis: Imaging, Lab Tests

Radiation (ionizing radiation) is used in hospitals and clinical labs to identify and treat health problems. Almost everyone knows of the use of x-ray machines to peer within the body and pin-point broken limbs, tumors, and enlarged organs. But few are aware of the other medical tools available. A process called "**scanning or imaging**" is used a great deal in hospitals throughout the US. Radioactive material (called radiopharmaceuticals) can be tagged or attached to the molecules of certain substances that, when injected into the human body, will concentrate in specific organs. A large detector (called a "gamma camera") is then placed over the patient to detect the locations where the specific substance, and hence the radioactive material, has concentrated. A two or three dimensional computer representation of the organ or organs will then show the physician areas that may indicate tumors or poorly functioning organs. If the liver were the organ of interest, then the physician might expect to see a uniform pattern displayed. The presence of lighter areas might indicate a tumor as the tissue of the tumor fails to absorb the tagging agent and the radioactive material—emitting less radiation. A darker area could indicate abnormally rapid absorption. In either case, areas of abnormal tissue can be more readily identified. Tests can be performed without exposing the patient to radiation. Blood samples are taken from the body and **tested in the laboratory** (in vitro) with techniques involving radioactive material.

Treatment (therapy):

After a successful diagnosis has been made, the patient can be *treated with radiation* to cure the disease, or at least reduce the problem. Radioactive material may be introduced into the body orally, such as using Iodine-131 to treat hyperthyroidism. As a concentrated mass sealed within a stainless steel encapsulation, radioactive material may be inserted directly into tumors in the body (brachytherapy) for a period of time. Radiation can be also used by directing a beam into the affected area or organ. Units containing large quantities of Cobalt-60 emit intense beams of gamma radiation (teletherapy) which can be directed into the area of the body needing treatment.

This technique minimizes the radiation exposure of healthy tissues. Accelerators (electronic/electrical devices that can accelerate particles in tight beams) are being used more and more each year.

Lasers, Cosmetic: Laser devices are rapidly expanding into the medical tool arena—although they emit a different form of radiation (non-ionizing radiation). Often, laser devices are used in conjunction with other radiation exposure devices to properly align radiation "beams" so that the tissues being exposed to a radiation beam are only those that need to be exposed. Lasers can be used to treat diseased tissues (such as polyp removal in the mouth and throat) and to allow rapid, low pain cosmetic treatments. Laser radiation generally presents a different type of radiation hazard than ionizing radiation.

Educational Uses: Teaching, Research

Many educational facilities (high schools, colleges, and training institutions) use radiation sources for *research* and training. Physics, chemistry, and biology labs have many radiation devices and procedures available. Radioactive materials are used to trace flow in systems, provide radiation beams for materials analysis, develop instruments, test systems, and to teach students the benefits and hazards of using radiation. Accelerators and x-ray devices are also used for materials analysis and other research. Electron microscopes emit low-energy x-rays in their operation. Many laser types, uses, and processes are researched and developed at educational facilities. Most of these tools are also used in the classroom and laboratories for *teaching* students. Since younger people are present and may be exposed, even greater care must be taken (the limits for radiation exposure to persons under 18 years of age are lower).

Industrial Uses

Non-Destructive Testing:

Industry also finds a great deal of use for radiation and radiation sources. Industrial radiographers use Iridium-192 and Cobalt-60 devices (usually called "cameras") to "x-ray" steel and other dense objects. This is generally called "*non-destructive testing*" because the tests are accomplished without changing the materials being tested. Most oil and gas pipeline welds are radiographed.

In well logging, instruments containing radioactive material and sensitive detectors are lowered into oil and gas wells to evaluate and *analyze* the geology under the surface and find crude oil, natural gas, and water formations.

Gauging: Measuring, Analysis
Radiation beams are emitted from *"nuclear gauges"*, passing through steel vessels and indicating or measuring fluid levels and densities electronically at remote locations—all without disturbing the contents of the vessels. Gauges (spinning pipe) can evaluate the lengths of used steel pipes that are intended to be re-used. This allows the identification of corrosion, erosion or other defects in the pipe before it can be placed into a system that, should even a single pipe fail, might cause devastating economic, environmental, and health problems. Gauges are also used to evaluate road bases and content during the construction of streets, highways, and parking lots. Some gauges are used to determine the structural integrity of older bridges and buildings. Others are used to analyze materials on surfaces—such as measuring the amount of lead in paint.

Sterilization: Medical Products, Food
Radiation can be used to *sterilize.* Large irradiator facilities contain millions of curies of radioactive material to irradiate sealed packages of medical and surgical supplies and instruments, thereby destroying all potentially infectious organisms within the package without harming the contents. Similar facilities are being used to irradiate **food** products—thereby greatly extending the shelf life of the products and virtually eliminating the possibility of "food poisoning". Although slightly changed, the food product is NOT radioactive—nor does it become toxic or unhealthful.

Power Generation:
Nuclear power plants provide a great deal of electric power with considerably less pollutants and waste than fossil-fueled processes. The nuclear plants' fuel (uranium) is mined by removing and processing ore from the ground, or by injecting fluids into the ground to free the uranium atoms and then removing the atoms chemically for later processing into fuel.

Miscellaneous Uses
Package X-ray, Light Sources, Smoke Detectors
There are many uses of radiation in industry that we take for granted. At airports, package x-ray units are used to x-ray baggage to prevent weapons and explosives from being taken onto aircraft. Many organizations use these devices to inspect incoming mail and packages that might contain explosives. Aircraft have lighted exit signs that do not require an electric source but will glow in the dark for years. Most smoke detectors use a small radiation source as a principle component of their sensor. Veterinarians can use many of the medical techniques previously discussed in the care of animals. The x-ray machine is an essential tool of the dentist.

Numbers of Users

These are a few highlights of the uses of radiation. We simply cannot appreciate the benefits to mankind presented by the many uses. The benefits have greatly outweighed the risks and harms. In 2000, the Texas Department of State Health Services regulated over 18,000 entities using radiation in Texas. Over 1,400 of these are licensed to use radioactive material at more than 2400 sites throughout the state. There are also many mobile operations conducted at other locations in the state every day.

In summary, working with radiation and radiation sources requires some knowledge and understanding of the tools and terms used in matters relating to radiation. We "enjoy" the benefits of using radiation to improve our quality of life, but we must remember that there are hazards associated with many of the uses and we must work carefully to prevent any serious events that may be harmful to ourselves and those around us.

Chapter 2

Introduction to Radiation Fundamentals

1. Introduction

This chapter will introduce basic radiation information terms and concepts. It is not intended to cover all possible aspects—simply the minimum information needed by a radiation safety officer to operate a radiation safety program or radiation worker in a "small or medium program" to protect himself or herself and to satisfy the minimum regulatory needs. There are many texts and government publications available to allow the reader to become more fully informed. Several are listed in the resources in the introduction of the book.

2. Characteristics of Radiation

The principle characteristics of radiation and radiation sources will be reviewed so that a better understanding of the safety systems employed for protection can be realized. Employers and government regulators will expect each radiation worker to follow all rules and safety procedures. By understanding some of the physical reasons for these safety procedures, we can better equip ourselves to abide by them and maintain a safe, non-threatening working environment. This section will introduce radiation and some of the sources of radiation.

Radiation Principles

Transfer of Energy

Radiation can be defined as the *transfer of energy* from one point to another point. An initial point, where a particular photon, particle, or other form of radiation, begins, would be the point of emission. A light bulb, for example, would emit a photon which travels through space until it encounters something (an atom). Thus, visible light is a form of radiation. In this example, the light bulb would be a "radiation source" and the atom (collection of atoms) would be "exposed to radiation". In general, a radiation source emits radiation which travels to and exposes an object—transferring energy from one point to another in the process.

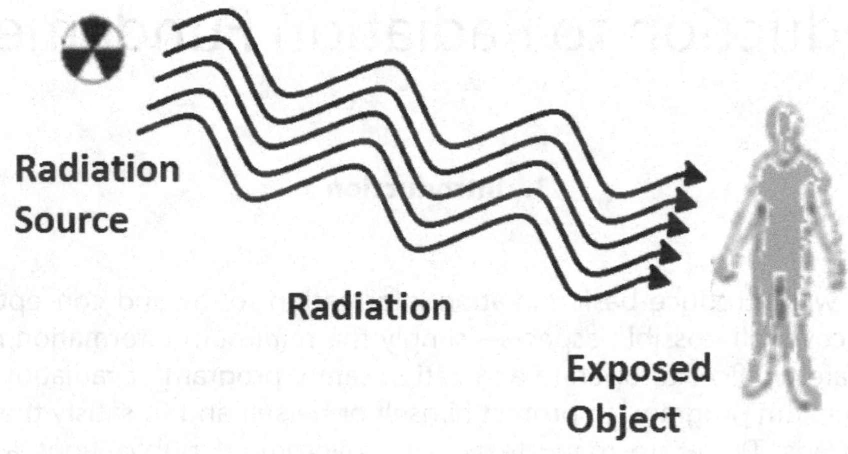

Figure 2-1 Radiation—Transfer of Energy

Particles or Waves: Electromagnetic Radiation

Radiation can occur in the form of *particles or waves*—and some types have characteristics of both. An analogy of particulate radiation might be a baseball thrown against a wall. The baseball would bounce off the wall at some angle and lower velocity—leaving a depression and warm spot on the wall. Energy was transferred from one point to another. The thrower represents the radiation source, the ball represents a particle (or radiation), and the wall is the object exposed. A water wave is an example of radiation in the form of a wave. An underwater earthquake imparts energy to water creating a pressure wave, which then travels through the body of water until it encounters an object and transfers some energy. Photons are *electromagnetic radiation*, although they demonstrate properties of both waves and particles and are often called "wavicles".

Ionizing and Non-ionizing Radiation

Radiation is generally classified as ionizing or non-ionizing. Radiation encountering objects (composed of atoms) can cause electrons in orbits around the objects' atoms to be knocked out of orbit. This leaves a positively charged atom and a free electron—or a positive and a negative "ion". The process is called ionization and the two ions are called an *ion pair*. Radiation with sufficient energy to cause ionizations is called ionizing radiation. Non-ionizing radiation does not cause ionizations. X-Ray radiation from machines is ionizing radiation while visible light is non-ionizing radiation. Ionizing radiation is capable of penetrating matter. Depending on the type of radiation, it can penetrate through even the densest of materials. This property makes some types of radiation useful for evaluating solid structures.

Note: Ionizing radiation will be the principle topic of this book.

Radioactivity

To understand radioactivity, we must first gain a little understanding of the structure and role of the atom.

The Atom and Its Structure

All matter is made up of atoms and each of the individual elements are composed of atoms which are unique to the particular element. Figure 2-2 shows the basic structure of the atom in the "classical planetary model" form. The nucleus can be viewed as a "sun" and the electrons as "circling planets"—like the solar system. The components of interest are protons, electrons, and neutrons. The center portion of the figure is the atoms's nucleus—which is composed of at least one **proton**. "Circling" around the nucleus is one or more **electrons.** The simplest atom has one proton and one electron. This is the element hydrogen. The nucleus has one proton with a positive charge balanced by the one electron with negative charge. The **neutron** is another particle found within the atom (in the nucleus), but it has a neutral charge and only adds mass to the atom. Since the protons and neutrons are found within the nucleus, they are often referred to as nucleons. Figure 2-2 illustrates the helium atom which has 2 protons and 2 neutrons in the nucleus and 2 orbital electrons.

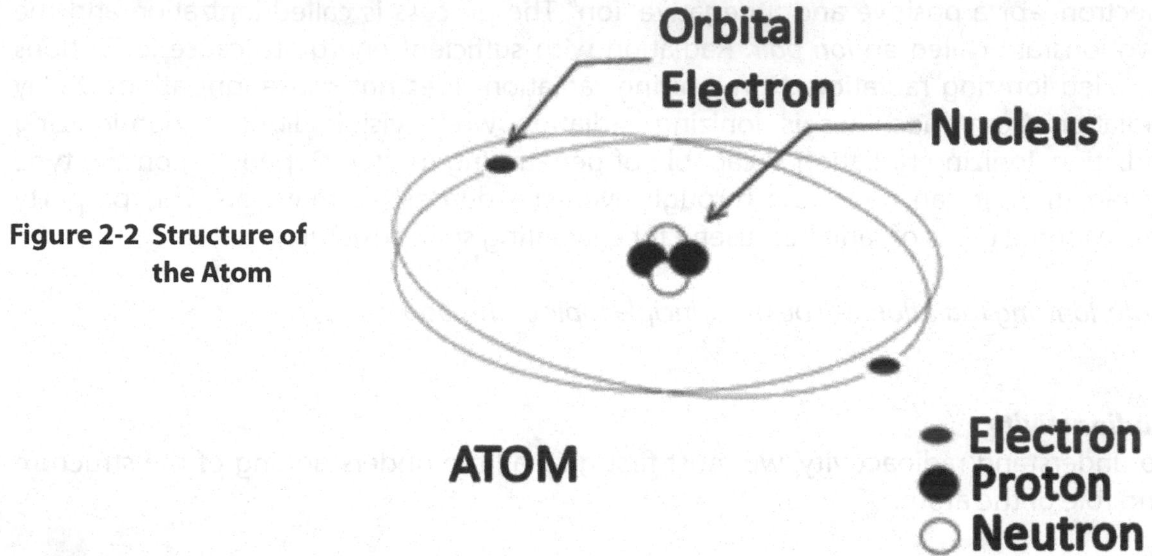

Figure 2-2 Structure of the Atom

Elements

As protons and electrons are added, new *elements* are created. Adding another proton, two neutrons, and another electron to the hydrogen atom would produce a helium atom. As long as the electrons and the protons are balanced, the atom remains neutral. The electrons are grouped according to their energies in "shells". The closer the electron is to the nucleus, the higher its energy (called binding energy). Should an atom have an extra proton or electron, then it is ionized. The chemical elements are identified according to the number of protons. The *atomic number (Z)* and the *mass number (A)* are used to identify the various elements. The atomic number is equal to the number of protons in the nucleus and the mass number is equal to the sum of protons and neutrons in the nucleus. For the helium atom shown above, the atomic number is 2 and the mass number is 4. It would be shown symbolically as ${}_2^4$He.

Isotopes

An atom with a different number of protons and neutrons in the nucleus is an isotope of its element. The isotopes have the same *chemical properties* as the element. When presenting isotopes, the Helium isotope with 2 additional neutrons (A = 4 + 2 = 6) would be shown as 6 He. The most common and popular form of expressing an isotope is by using the form He-6. *In this document, the terms isotope, radioisotope, and radioactive material will be used interchangeably.*

Radioactive Quantity

Curie, Becquerel

Now that we know what a radioactive isotope is, how do we quantify it? We define a unit of radioactivity (at the time of determination) by the number of disintegrations over time, as:

1 Curie equals 3.7 X 10^{10} disintegrations per second.

1 Curie is approximately the total radioactivity of 1 gram of Radium-226 and was named after Marie and Pierre Curie, early researchers in radiation. This unit of radioactivity was defined many years ago and is called a "traditional unit". Now, under the Systeme International d'Unites (SI), the unit of radiation is defined as:

1 Becquerel equals 1 disintegration per second.

Both units are currently used, however the Curie is slowly being phased out. In this book, units of Curie will be used, parenthetically accompanied by the SI equivalent.

Unit conversions relating to Becquerels and curies are:

1 picocurie (pCi)	equals	37 millibecquerel (mBq)
1 nanocurie (nCi)	equals	37 Becquerel (Bq)
1 microcurie (µCi)	equals	37 kilobecquerel (kBq)
1 millicurie (mCi)	equals	37 megabecquerel (MBq)
1 Curie (Ci)	equals	37 gigabecquerel (GBq)
1 kilocurie (KCi)	equals	37 terabecquerel (TBq)

1 Becquerel (Bq)	equals	27 picocuries pCi
1 kilobecquerel (kBq)	equals	27 nanocuries (nCi)
1 megabecquerel (MBq)	equals	27 microcuries (µCi)
1 gigabecquerel (GBq)	equals	27 millicuries (mCi)
1 terabecquerel (TBq)	equals	27 Curies Ci

Radioactive Decay

The atoms of some elements have *excess energy* and are *unstable* (their nuclei are unstable). Through a process called *radioactive decay* (or disintegration), such an atom will transform itself into another atom by spontaneously emitting energy and particles (radiation). We call elements composed of these atoms *radionuclides*. If the element's atoms are isotopes, then they are *radioisotopes*. The atoms of a radioisotope will make these transformations at a uniform, decreasing rate—not all at once or in groups. Beginning with a given quantity of a radioisotope, after a period of time, a number of the atoms will have transformed into stable atoms, so the remaining atoms are the ones contributing to the rate of decay. This causes a decrease in the overall decay rate over time. To quantify the rate of decay, we introduce the concept of **half-life**. The half-life is the amount of time required for a given isotope to decay to a quantity that is one half of its original quantity. For example, an isotope with a half-life of one day would have ½ of the original number of unstable atoms after one day. After two days, there would be ½ of ½, or ¼. Day three, ½ of ¼, or ⅛. Day four, ½ of ⅛, or 1⁄16, and so on . . Figure 2-3 shows a plot of the radioactivity decreasing over time due to decay. Of great importance is the fact that different isotopes have different, specific rates of decay.

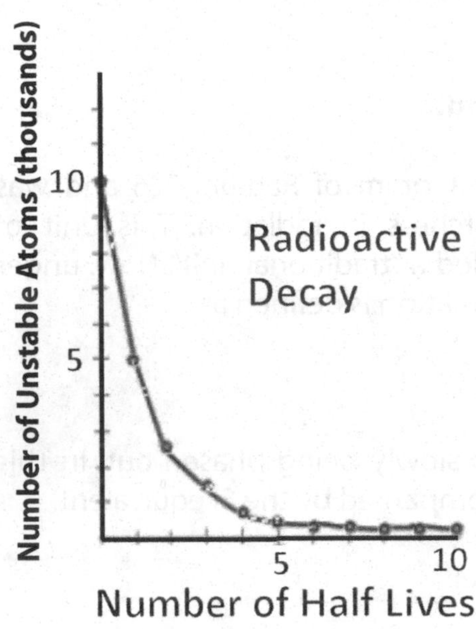

Figure 2-3 Decay of Unstable Atoms Over Time.

For example, if we had one Curie of an isotope today and we measured the number of disintegrations we would determine 3.7×10^{10} disintegrations per second (dps).

- If the isotope had a half-life of one day, at this same time tomorrow we would measure 1.85×10^{10} dps (one half of the unstable atoms would have transformed and there would be one half of a Curie remaining).

- If the isotope had a half-life of 1,000 years, at this same time tomorrow we would still measure 3.7×10^{10} dps because so few of the unstable atoms would have transformed.

Origins of Radiation Sources

Natural, Manmade

Ionizing radiation can be emitted by either natural or man-made sources. Natural sources are composed of radioactive atoms that were formed when the universe was created (or atoms that have resulted from the decay of those atoms). Man-made sources are devices which electronically produce radiation or are atoms made radioactive by man.

Natural Sources: Uranium, Cosmic Rays

The most common source of *natural radioactivity* is due to the presence of *uranium* in the Earth's soils and geological structures. Thought to have been produced during creation of the universe, the uranium atom (which has 3 natural isotopes) is very heavy and unstable. When it decays, it produces an alpha particle and another heavy atom (called a daughter) which is also unstable. The atoms produced vary, so that eventually, a mixture of heavy atoms of uranium and isotopes of other elements become distributed throughout the original "mass". Eventually, the initial collection of atoms becomes a collection of isotopes of a number of elements. For example, uranium-238 (with a half-life of 4.5 billion years) emits an alpha particle (there are 3 possible energies) as the atom decays to thorium-234. This atom (half-life of 24 days) in turn decays to palladium-234 while emitting a beta particle (of three possible energies). The palladium atom (half-life of 6-7 hours) decays by emitting a beta particle and becomes uranium-234. And so on . . . Now since all of these have differing half-lives, one can see that there will be a mixture of isotopes in the collection. This process is called a "*decay series*". While there are many decay series, only a few occur naturally.

Another source of natural radiation is atoms made unstable by neutrons released from cosmic rays coming from space and interacting with matter on the Earth. Most cosmic rays (of unknown origin) are screened out by the Earth's magnetic fields and atmosphere. The high energy cosmic rays that do penetrate through and interact sometimes produce neutrons—which can be captured by the nucleus of an atom and cause the atom to become unstable. Naturally occurring Carbon-14 and Potassium-40 are examples of this process.

Man-made Sources: Use of Natural Radioactivity, Electrical Devices

There are two methods used by man to produce radiation and radiation sources—using naturally occurring radioactivity or using electrical devices.

One method of producing radioactive material is to use existing *natural radioactivity* through the "fission" process. When a neutron interacts with a uranium atom, the atom can be split into two "roughly" equal parts thereby creating two atoms of other elements and releasing energy—including neutrons. If the "released" neutrons interact with other atoms, they can change them to isotopes of other elements. This provides a tool for creating many isotopes. Nuclear reactors, using uranium or plutonium as fuel, produce a high neutron flux. Exposure of various elements to the high flux allows us to produce specific, useful isotopes—a process called *neutron activation*. Neutron activation can be accomplished through other means. It should be noted that an alpha particle interacting with certain types of atoms (elements) can produce neutrons. Mixing plutonium-239 with stable beryllium, for example, allows the alphas released from the decaying plutonium atoms to interact with the nuclei of the beryllium atoms—releasing a neutron in the process. Neutrons of a given energy range can be made to interact with atoms of other elements, creating various isotopes. One neutron source produced in reactors, Californium-252, releases neutrons in a sufficiently high flux that it can be used for neutron activation at a practical level.

If we can produce high-energy neutrons without the use of heavy atoms, we could rather cleanly produce useful radioactive materials. The use of *accelerators* allows us to do this. Simply, an accelerator is a device that moves a charged particle (ion) by application of an electric potential or electromagnetic field (usually both) so that the particle increases in energy or participates in an interaction with matter creating other particles of higher energy than the original particle. If we produce neutrons then we can use neutron activation to form isotopes. Some accelerators are operated in a line (linear accelerator) and others use a circular path to increase the energy potential. By directing the final particles into a target of atoms, the particles emitted from the resulting interactions can be studied to analyze atomic structure. Accelerators can also be used for medical treatment and to produce radioactive material.

The x-ray machine is an electrical device that produces useful radiation (the method will be discussed later in this chapter).

2. Properties of Radiation

Having reviewed the various sources of radiation, we now need to examine radiation and how it affects us. There are a number of different types and forms of radiation. Each has its own properties.

Types

Ionizing and Non-ionizing

As previously indicated, the two primary forms of radiation are ionizing and non-ionizing. However, ionizing radiation is the primary focus of this text.

Forms of Radiation

Particulate, Electromagnetic

The two forms of ionizing radiation are particulate and electromagnetic.

Particulate radiation consists of alpha, beta, and neutron particles. There are others, but these are the ones that are commonly used or encountered in dealing with radiation sources. Alpha, beta, and/or neutron particles are emitted during the decay of unstable atoms. Note that a given unstable atom does not emit either an alpha or beta during decay, but the decay of different isotopes determines the particular emission. For example, an americium-241 atom decays releasing an alpha particle (alpha decay) while an americium-242 atom decays emitting a beta particle (beta decay). The *alpha particle* is a large particle consisting of two protons and two neutrons all bound together. Essentially, it is the same as a helium atom stripped of its two electrons. The *beta particle* is essentially a high-speed electron and cannot be physically distinguished from an electron. *Neutrons* are composed of a proton and an electron bound together. They are produced through the interactions previously discussed. Neutron decay (decay of the neutron itself) produces a proton and a beta particle.

Electromagnetic (photonic) radiation includes two forms that we need to consider: *Gamma Rays and X-Rays*. Gamma rays and x-rays are physically indistinguishable and are identified by their method of production. If they come from the nucleus of an atom, they are gammas. If they are produced by an interaction with an orbital electron of an atom, then they are x-rays (discussed later in this section). Decay of an unstable atom releasing a Gamma is "gamma decay". Gammas and x-rays are electromagnetic waves.

Properties of Radiation

Each of the forms of radiation described above has properties and characteristics that make them useful for specific functions. The properties must also be considered when dealing with radiation protection.

Particle Size

The space within an atom is quite empty. If the nucleus of an atom were the size of a softball, then the nearest electron would be orbiting several miles away. Thus, the relative sizes of the different types of particles are important since the size affects the probability of interaction and also affects the particulate ranges and the types of interactions:

Particle	Symbol	Size
Alpha	α	very large particle
Neutron	η	large particle
Beta	β	small particle
Gamma	γ	negligible size
x-ray	γ	negligible size

Table 2-1 Comparison of Particle Sizes

Before proceeding further, we need to note the following definitions.

Definitions:

electron volt (eV)	the unit used to express the energies of particles and photons of radiation. One (1) electron volt is the energy equivalent of an electron with electrical potential of 1 volt. Any other particle or photon of 1 eV has the same amount of energy.
Kiloelectron volt (KeV)	1000 electron volts (10^3 eV).
Megaelectron volt (MeV)	1,000,000 electron volts (10^6 eV).

Energies

The *energy range* of each form is an important consideration in examining their properties. The approximate energy ranges for each are:

Particle		Energy Range
alpha	<	4-7 MeV
beta	<	0-7 MeV
neutron	<	0->10 MeV
gamma	<	0-5 MeV
x-ray	<	0-10 MeV

Table 2-2 Comparison of Energy Ranges

Range of Travel in Air

The ranges that the particles or photons travel depends on a number of factors: media (such as air), energy, charge, and particle size. For comparison, the maximum travel ranges-in air-for each are:

Particle		Range in Air
alpha	<	10 cm
beta	<	0-10 m
neutron	<	over 37 Km[2]
gamma	<	0-100 m
x-ray	<	0-100 m

Table 2-3 Comparison of Particle Travel Ranges in Air

Shielding

Shielding from radiation will be discussed later in this study, so a brief comparison will be considered here. The materials used for shielding vary for the different

2 Estimated by half-life of 11.7 secs, average life of 1.443 times half-life, and velocity of 2200 m/sec for fast neutrons.

particles and photons. Usually, density of the shielding material determines its effectiveness as a shield.

The following table indicates some of the shielding materials that are effective:

Particle		Shielding
alpha	<	any dense material (a sheet of typing paper will stop most)
beta	<	moderately dense materials (such as plexiglass)[3]
neutron	<	materials with many hydrogen atoms[4]
gamma	<	dense materials (such as lead)
x-ray	<	dense materials (such as lead)

Table 2-4 Shielding Materials for Various Radiations

X-Ray Generation and Characteristics

Since many operations are conducted using x-ray equipment, the radiation safety officer, or person responsible for radiation safety, should be familiar with the generation of x-rays and their characteristics.

X-ray Production: Bremsstrahlung, Characteristic

X-Rays are one type of man-made radiation. If an electron passes sufficiently near the nucleus of an atom (missing the orbital electrons), it can leave the vicinity of the nucleus with its course changed and its energy reduced (see Figure 2-4). When this happens, energy is emitted in the form of an x-ray photon (electromagnetic energy). The resulting x-ray is called *"Bremsstrahlung"* radiation. If the "incoming" electron interacts

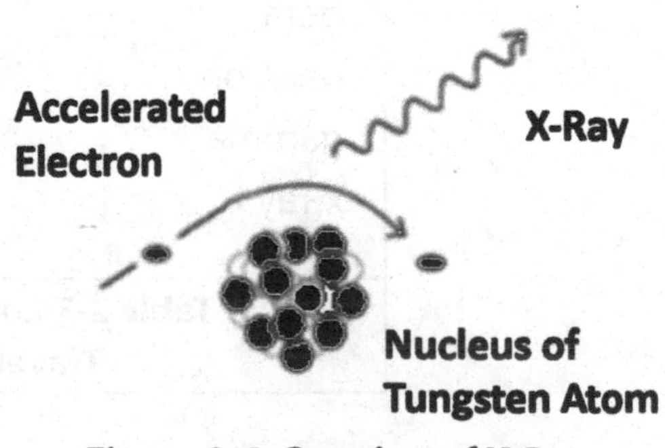

Figure 2-4 Creation of X-Ray

---

3 Dense materials, such as lead, have many atoms capable of interacting with betas and will produce a large amount of Bremsstrahlung radiation (see next section). Less dense materials can be used to shield for the betas and then denser materials used to shield for the resulting Bremsstrahlung.

4 Neutrons interact well with hydrogen atoms—less well with larger atoms.

with an inner shell electron of an atom by ejecting it out of the atoms structure, outer orbit electrons can drop into the "vacant position". When this happens, a *characteristic x-ray* is emitted from the atom. Bremsstrahlung x-rays have a broad range of energies while characteristic x-rays have a narrow range of energies.

X-Ray Generation: Electron Beam, Target, X-Ray Tube

Like gamma rays, x-rays are capable of penetrating matter. The higher the energy of the x-rays, the greater the depth of penetration for a given material. In order to use this phenomenon to produce useful x-rays, an *electron beam* (see Figure 2-5) is directed from a cathode toward an anode, or target, composed of a mass of atoms in a desirable arrangement. The target (usually tungsten as solid metal) is placed in a glass enclosure along with a heated filament for an electron source and the air is removed. A positive high voltage is placed on the target to accelerate the electrons formed around the heated (negatively charged) filament towards the (positively charged) target. As the electrons collide with the atoms of the target and interact with them, x-rays are emitted. We can construct an "x-ray tube" by placing shielding material (material that few x-rays can penetrate) around the glass enclosure—except in one area. X-rays will then emanate from the tube only in the unshielded area. Thus, an x-ray beam is produced and can be directed toward an object. The intensity of the x-ray "beam" lessens as the distance between the x-ray source and the measuring point increases (just as a flashlight appears dimmer and dimmer as the observer moves farther and farther away). This concept will be covered in more detail. A filter is usually placed in the beam to absorb low energy X-rays that are not useful, and a collimator is generally used to restrict the width of the beam. These help to reduce the emission of unnecessary x-rays that can cause unwanted or unintended exposure.

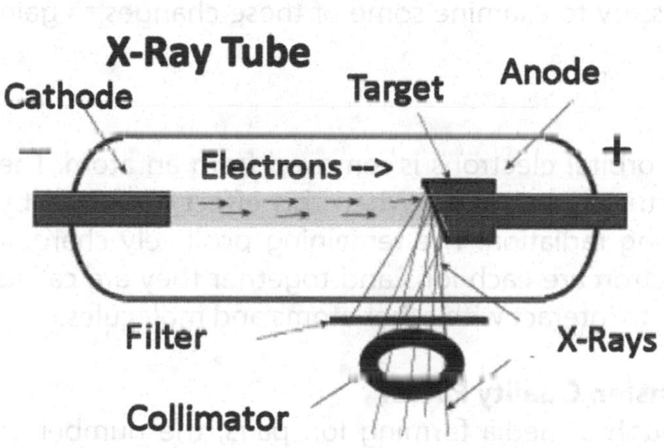

Figure 2-5 Production of X-Rays-X-Ray Tube

X-Ray Effects: Scatter or Secondary Radiation, Primary Radiation

As previously indicated, X-rays encountering and reacting with the atoms of an object cause ionizations and produce ion pairs—continuing on at lower energy levels and on different paths. Thus, if an x-ray beam is directed towards a target, interacting x-rays of the beam can be re-emitted (at lower energies) from the object in many directions. This is called *scatter* or *secondary radiation*. The same phenomenon occurs when gammas are being used, especially as a beam. Air, which is between the x-ray tube and the object for most applications, is a major contributor of scatter radiation. The original beam is called *primary radiation*.

General Radiation Effects

When radiation (whether particulate or photonic) encounters an object, such as an atom, energy is transferred to the object and both the initial form of radiation and the object experience changes. It is necessary to examine some of these changes to gain an understanding of radiation effects.

Ionization Process: Ion Pair

Ionization occurs when one or more orbital electrons is removed from an atom. The atom (or molecule) is no longer electrically balanced. When this effect is caused by a particle or photon, we call it ionizing radiation. The remaining positively charged photon and (negatively charged) electron are each ions and together they are called an "*ion pair*". The ions formed are "free" to interact with other atoms and molecules.

Ionization Effect: Linear Energy Transfer, Quality Factor

When ionizing radiation passes through a media forming ion pairs, the number of ion pairs formed along the path per linear centimeter is called specific ionization. If the media should be water or tissue, the transfer of energy per unit distance is *Linear Energy Transfer* (LET). The units used are kiloelectron volt (keV) per micrometer (μm). The effect of a given type of radiation and energy on a volume of air is different than the effect on an equal volume of water or human tissue. We need to have a system of relating the radiation measured in air (where most measurements are made) to the effect that radiation will have on tissue. As LET increases, the biological effect caused by the ionizing radiation increases. We use the LET and the *Quality Factor (Q)* to estimate the exposure and effect when tissue is irradiated. The Quality Factor is a unitless modifier which increases as LET increases.

Type of Radiation	Quality Factor (Q)	Absorbed Dose Equal to a Unit Dose Equivalent[5]
X, gamma, or beta radiation and high-speed electrons	1	1
Alpha particles, multiple charged particles, fission fragments and heavy particles of unknown charge	20	0.05
Neutrons of unknown energy	10	0.1
High-energy protons	10	0.1

"Absorbed dose in gray equal to 1 Sv or the absorbed dose in rad equal to 1 rem.

Table 2-5 Mean Quality Factors and Absorbed Dose Equivalencies[5]

For monoenergetic neutrons, mean quality factors and fluence for dose equivalents, as listed in given in TAC §289.201(n)(2):

Table 2-6 Mean Quality Factors, Q, and Fluence Per Unit Dose Equivalent for Monoenergetic Neutrons[6]

	Neutron Energy (MeV)	Quality Factor** (Q)	Fluence per Unit Dose Equivalent* (neutrons cm-2rem-1)	Fluence per Unit Dose Equivalent* (neutrons cm-2Sv-1)
(thermal)	2.5 x 10-8	2	980	980
	1.0 x 10-7	2	980	980
	1.0 x 10-6	2	810	810
	1.0 x 10-5	2	810	810
	1.0 x 10-4	2	840	840
	1.0 x 10-3	2	980	980
	1.0 x 10-2	2.5	1,010	1,010

5 TAC §289.201(n)(2)

1.0 x 10-1	7.5	170	170
5.0 x 10-1	11	39	39
1.0	11	27	27
2.5	9	29	29
5.0	8	23	23
7.0	7	24	24
10	6.5	24	24
14	7.5	17	17
20	8	16	16
40	7	14	14
60	5.5	16	16
1.0 x 102	4	20	20
2.0 x 102	3.5	19	19
3.0 x 102	3.5	16	16
4.0 x 102	3.5	14	14

*Monoenergetic neutrons incident normally on a 30-centimeter diameter cylinder tissue equivalent phantom.

** Value of quality factor (Q) at the point where the DE is maximum in a 30-centimeter diameter cylinder tissue-equivalent phantom.

Application of these values will become more apparent after discussion of radiation units in the next section.

Radiation Units

Measurement of Ionizations

Now that we know what radiation is, we must be able to measure it in order to make use of it and to compare its effects on various types of matter. Both the "traditional" and SI units will be addressed.

Radiation can be measured using the number of ion pairs created. The simplest method is to determine (electrically) the number of ion pairs created in a volume of air. A chamber, enclosing a volume of air using materials that do not absorb very much radiation, is used. Two charged "plates" are located within—one positive and one negative. Radiation penetrating the chamber creates ion pairs in the air—also of positive and negative charge. The ions with positive charge are attracted to the negative plate and the ions with negative charge are attracted to the positive plate. This causes a measurable change in the voltage between the plates. As the energy and/or the intensity (number of x-rays) increases, the voltage change increases. The chamber is an *ionization chamber* (shown in Figure 3-5).

Exposure: Roentgen

The *Roentgen* (symbol "R" or "r") is a traditional unit of measurement of radiation. It is the amount of ionizing radiation (radiation flux) that will produce 2.08 billion ion pairs in one cubic centimeter at Astandard temperature and pressure. This is also equivalent to 84 ergs (erg is a unit of work). New definitions and units of measurement have been instituted, and are slowly being implemented, for international standards, but this discussion will deal primarily with the traditional standards, which can still be used. If radiation interacts with materials other than air, then the energy deposited will be different. One cubic centimeter of water will have a different number of ion pairs produced when exposed to the same amount of radiation as one cubic centimeter of air. The same is true for irradiation of tissue. The effect on tissue is very nearly the same as for air and water—so simple measurements often allow the assessment of the amount of exposure of human tissue. But this holds only for certain forms and energies of radiation. For others, we consider the Quality Factor previously discussed.

Dose: rad, gray, rem, Sievert

The amount of radiation absorbed by a material is called *dose*. Dose takes into consideration the specific type of radiation involved and the relative effects of that radiation on the object exposed. A unit of measurement, the *RAD* (Radiation Absorbed Dose), has been defined as the amount of radiation causing a material to absorb 100 ergs per gram of material. The SI unit of absorbed dose is the gray (gy), where one gray is equal to an absorbed dose of 100 rads.

1 gray = 100 rad.

Dose Equivalent

We are most interested in the effect of radiation absorbed by human tissue—so we use the term *Relative Biological Effect*, or *RBE*. The *Rad Equivalent Man,* rem or REM, takes this into account and is defined as the dose in rads times the RBE. The SI unit of dose (equivalent) is the Sievert (Sv), where one Sievert equals one hundred rem.

1 Sv = 100 rem (100,000 mrem)

0.01 Sv = 1 rem (1000 mrem)

It is defined in the regulations as: the dose equivalent in sievert is equal to the absorbed dose in gray multiplied by the quality factor (1 Sv = 100 rem). The traditional units will be the most used in this discussion with the equivalent SI units parenthetically introduced. For most of the energies of radiation we encounter, the dose in rem is approximately equivalent to the dose in rads and to the dose in r. It is common to use the r and the rem interchangeably.

Example:

A worker received an exposure from a radiation field of 1 r per hour. What would be the dose if the worker was in the area for 2 hours?

1 r/hr X 2 hrs = 2 r (exposure) or a dose of 2 rem (assumes a Q of 1).

What is the dose in Sieverts if the exposure was to neutrons of unknown energy?

2 rem X 10 = 20 rem (quality factor of 10 for neutrons of unknown energy)
20 rem = 20/100 = 0.2 Sv

Radiation Levels

For gamma and x-ray radiation, the radiation level caused by a radiation source does not remain the same in all areas because the level decreases as distance from the source increases. Nor is the radiation level from one isotope the same as the radiation level from a different isotope of the same quantity.

Radiation Level Changes: Intensity, Inverse Square Law

If one measures the radiation level (*intensity*) at a given distance from a point source (a discrete source with small dimensions) and then measures the radiation level again at a point of greater distance, the latter will be lower. (This is similar to the effect of moving away from a burning light bulb and observing that the light appears to become dimmer and dimmer.) The intensity decreases proportionately as distance increases. In fact, the intensity is inversely proportional to the square of the distance from the source. This is called the *Inverse Square Law*.

Inverse Square Law

The inverse square law can be mathematically expressed as:

where

I_1 is the intensity at a distance d_1 from the source and
I_2 is the intensity at a distance d_2 from the source.

$$\frac{I_1}{I_2} = \left(\frac{d_2}{d_1}\right)^2$$

Figure 2-6 Inverse Square Law

For a given radiation source, if we know I_1 at a distance d_1, then we can calculate I_2 at a distance of d_2. We algebraically manipulate the above equation to:

$$I_2 = I_1(d_1/d_2)^2$$

The parameters are shown graphically in Figure 2-7.

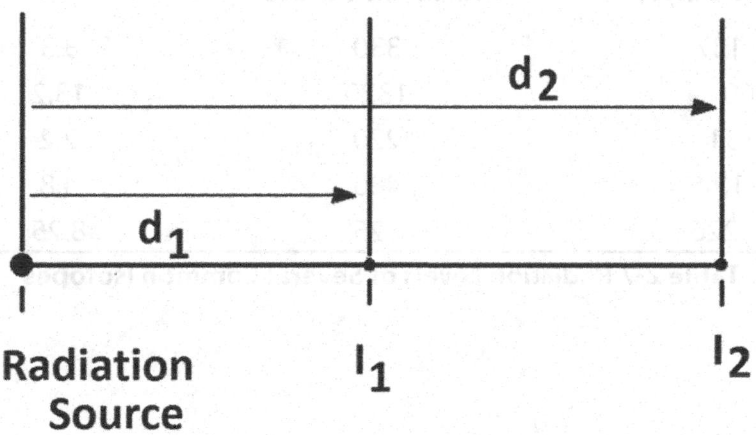

Figure 2-7 Graphic Representation of Inverse Square Law

Example: a Cesium-137 source of 1 Curie has a radiation level of .33 R/hr at a distance of 1 meter (gamma ray constant). We want to determine the radiation level at 2 meters.

$I_1 = 0.33$ R/hr; $d_1 = 1$ m; and $d_2 = 2$ m

Using the above equation and substituting:

$$I_2 = I_1 (d_1 / d_2)^2$$
$$I_2 = (0.33) X (2)^2 = (0.33) X (3) = 0.0825$$

Thus $I_2 = 0.0825$ R/hr (0.000825 Sv/hr) at 2 meters from the source. This can also be expressed as 82.5 mr/hr (0.825 mSv/hr).

This relationship is frequently used in radiation safety practice. A "rule of thumb" falls from this relationship: Doubling the distance reduces the radiation level to one-fourth.

Radioisotopes

Radioisotopes have different radiation levels when compared at the same concentrations. As an example, the following are the radiation levels for some common radioisotopes:

Radioisotope (1 Curie)	Radiation Level (mr/hr) at 1 meter	Radiation Level (mSv/hr) at 1 meter
Cesium-137	330	3.3
Cobalt-60	1320	13.2
Iodine-131	220	2.2
Iridium-192	480	4.8
Radium-226	825	8.25

Table 2-7 Radiation Levels of Several Common Isotopes

Summary of Radiation Units:

Unit	Traditional (Symbol)	SI (Symbol)
Radioactivity (quantity)	Curie (Ci)	Becquerel (Bq)
Exposure (radiation)	Roentgen (R)	None
Absorbed Dose	RAD	Gray (Gy)
Dose equivalent (man)	rem	Sievert (Sv)

Table 2-8 Summary of Radiation Units

3. Significance of Radiation Dose

For radiation sources that are totally outside of the body (sealed sources of radioactive material, x-ray devices, etc.), radiation exposure of humans can only consist of the exposure of the body and/or one or more extremities. Using traditional terms and concepts, a person standing next to a x-ray device during operation might be exposed to the radiation being emitted by the system—thereby resulting in a "whole body" dose. A person placing a hand into a radiation "beam" would have an exposure to the "extremity".

On the other hand, a person that inhales, ingests, absorbs through the skin, or has radioactive material injected into the body, will experience internal exposure. When radioactive material gets into the body, it may accumulate in one or more organs, causing an organ dose. If this occurs, the organ will become a Aradiation source, exposing not only the organ itself, but also (depending on the type of radioactive material and chemical form) other organs and the whole body.

Government regulations limit doses to the body and to individual organs. They also require that each company/facility using radiation sources conduct operations such that radiation exposures of workers (and anyone else) be "As Low As Reasonably Achievable"—or ALARA. The dose limits are designed to take the sum of all exposures into account—from both external and internal exposures. The following (Table 2-9) is a summary of terms related to exposure and dose and defined in the current regulations (although only a few of them are commonly used).

Regulatory Dose Definitions

Before reviewing the regulatory limits regarding exposure, we should be familiar with the following terms and definitions used in regulations[6]:

Table 2-9 Regulatory Exposure and Dose Definitions

Occupational Dose	The dose received by an individual in the course of employment in which the individual's assigned duties involve exposure to sources of radiation from licensed/registered and unlicensed/unregistered sources of radiation, whether in the possession of the licensee/registrant or other person. Occupational dose does not include dose received from background radiation, from any medical administration the individual has received, from exposure to individuals administered radioactive material and released in accordance with this chapter, from voluntary participation in medical research programs, or as a member of the public.
Public Dose	The dose received by a member of the public from exposure to sources of radiation released by a licensee, or to any other source of radiation under the control of a licensee/registrant. It does not include occupational dose or doses received from background radiation, from any medical administration the individual has received, from exposure to individuals administered radioactive material and released in accordance with this chapter, or from voluntary participation in medical research programs.

6 As listed in *25 TAC §289.201(b)*

Table 2-9 Regulatory Exposure and Dose Definitions

Exposure	The quotient of dQ by dm where "dQ" is the absolute value of the total charge of the ions of one sign produced in air when all the electrons (negatrons and positrons) liberated by photons in a volume element of air having mass "dm" are completely stopped in air. The SI unit of exposure is the coulomb per kilogram (C/kg). The roentgen is the special unit of exposure. For purposes of this chapter, this term is used as a noun.
Dose	A generic term that means absorbed dose, dose equivalent, effective dose equivalent, committed dose equivalent, committed effective dose equivalent, total organ dose equivalent, or total effective dose equivalent (in the rules, "radiation dose" is an equivalent term).
External Dose	That portion of the dose equivalent received from any source of radiation outside the body.
Internal Dose	That portion of the dose equivalent received from radioactive material taken into the body.
Absorbed Dose	The energy imparted by ionizing radiation per unit mass of irradiated material. The units of absorbed dose are the gray (Gy) and the rad.
Dose Equivalent	The product of the absorbed dose in tissue, quality factor, and all other necessary modifying factors at the location of interest. The units of dose equivalent are the sievert (Sv) and rem.
Effective Dose Equivalent	The sum of the products of the dose equivalent to each organ or tissue (HT) and the weighting factor (wT) applicable to each of the body organs or tissues that are irradiated (HE = ΣwTHT).

Table 2-9 Regulatory Exposure and Dose Definitions

Committed Dose Equivalent	The dose equivalent to organs or tissues of reference (T) that will be received from an intake of radioactive material by an individual during the 50-year period following the intake.
Committed Effective Dose Equivalent (CEDE)	The sum of the products of the weighting factors applicable to each of the body organs or tissues that are irradiated and the committed dose equivalent to each of these organs or tissues (HE,50 = ΣwT, HT,50).
Total Effective Dose Equivalent (TEDE)	The sum of the effective dose equivalent for external exposures and the committed effective dose equivalent for internal exposures.
Total Organ Dose Equivalent (TODE)	The sum of the deep dose equivalent and the committed dose equivalent to the organ receiving the highest dose as described in the rules.
Deep Dose Equivalent	(applies to external whole body exposure) The dose equivalent at a tissue depth of 1 centimeter (cm) (1,000 milligrams per square centimeter (mg/cm2)).
Shallow Dose Equivalent	(that applies to the external exposure of the skin of the whole body or the skin of an extremity)—The dose equivalent at a tissue depth of 0.007 cm (7 mg/cm2).
Lens Dose Equivalent	The external dose equivalent to the lens of the eye at a tissue depth of 0.3 cm (300 mg/cm2).
Quality Factor (Q)	The modifying factor used to derive dose equivalent from absorbed dose.

In most cases, the TEDE will be the type of dose that will be dealt with and the radiation safety officer or person responsible for radiation safety will be responsible for making the assessment and complying with the requirements. Those with unsealed sources may need to deal with the TODE. Beyond that, a specialist in dose calculation and assessment will probably be needed to deal with more complex exposures—which, fortunately, are rather rare. The more complex dose assessments are beyond the scope of this text

Types of Exposure

Exposure: External, Internal

External exposure is exposure caused by radiation sources OUTSIDE of the body. The exposure can be from a sealed source of radioactive material, contaminated items, radiation machine devices, and gaseous or dust clouds. The current regulatory limits are summarized below. The exposure (dose equivalent) of an individual from external sources is relatively easy to determine.

Internal exposure is exposure caused by radioactive material INSIDE of the body. This type of exposure (dose equivalent) is usually rather difficult to assess. The manner that the radioactive material entered the body, the physical and chemical forms of the radioactive material, and the tendency for the various organs of the body to retain or eliminate the material all affect the determination of the internal dose(s).

Total Dose: Summation, Body And Internal Organs, Extremities, Lens Of Eye

The above *exposures are summed* to evaluate the total dose equivalent of any organ so that they can be compared to regulatory limits. The lens of the eye, the skin, and the extremities are not included as they have separate limits.

Dose Limits for Radiation Workers

The dose limits for radiation workers in Texas (which will be the same for all NRC and Agreement States) are specified as:

An annual limit shall be the more limiting of:

- the total effective dose equivalent being equal to 5 rems (0.05 Sv); or
- the sum of the deep dose equivalent and the committed dose equivalent to any individual organ or tissue other than the lens of the eye being equal to 50 rems (0.5 Sv).

The annual limits to the lens of the eye, to the skin of the whole body, and to the skin of the extremities shall be:

- a lens dose equivalent of 15 rems (0.15 Sv); and
- a shallow dose equivalent of 50 rems (0.5 Sv) to the skin of the whole body or to the skin of any extremity.

Table 2-10 Radiation Worker Exposure Limits[7]

Thus, a radiation worker can only receive a maximum of 5,000 millirem in a year for a total effective dose equivalent (exposure to the entire body). Individual doses to the different body organs, lens of the eye, skin (i.e., shallow dose equivalent), and to extremities can be higher—up to 50 rem.

Radiation Limits

Restricted areas, Unrestricted areas

The principle of ALARA, discussed above, applies to ALL persons that may be exposed to radiation—even members of the general public. In order to protect the public health and safety, steps must be taken to control exposures of all persons that may be potentially affected by the use of radiation.

Regulations define a restricted area[8] as: "An area, access to which is limited by the licensee for the purpose of protecting individuals against undue risks from exposure to sources of radiation. Restricted area does not include areas used as residential quarters, but separate rooms in a residential building may be set apart as a restricted area." Since the area is controlled, persons entering can be monitored or controlled to assure that their exposure(s) does not exceed any limit. Occupationally exposed individuals are permitted to have higher exposures than individuals non-occupationally exposed. If a person's duties expect work with radiation, then it is anticipated that they will receive radiation exposures—but under a controlled situation. For all other areas, lower limits are provided by rule. Non-occupationally exposed persons can be prevented from entering restricted areas at times when they might be exposed. However, outside of the restricted areas, there are no controls. Recognizing that radiation levels cannot always be completely zeroed at the perimeter of a restricted area, limits have been set for radiation levels in "unrestricted areas". Regulations define *unrestricted area*[9] (uncontrolled area) as an area, or access to, which

7 25 TAC §289.202(f)

8 25 TAC §289.201(b)(90)

9 25 TAC §289.201(b)(118)

is neither limited nor controlled by the licensee. For regulatory purposes, "uncontrolled area" is an equivalent term.

Dose Limits for Individual Members of the Public (TEDE)

The dose limits for members of the general public are more stringent than those for radiation workers.

Radiation control rules[10] limit the exposure of members of the general public as follows:

A licensee or registrant must conduct operations so that

- the total effective dose equivalent (TEDE) to individual members of the public from the licensed and/or registered operation may not exceed

  0.1 rem (1 mSv) in a year[A]

- the total effective dose equivalent (TEDE) to individual members of the public from exposure to radiation from radiation machines[11] may not exceed

  0.5 rem (5 mSv) in a year

- the dose in any unrestricted area from licensed and/or registered external sources may not exceed

  0.002 rem (0.02 mSv) in any one hour

  [A]Exceptions: TEDE does not include the dose from disposal of radioactive material into sanitary sewerage, background radiation, exposure of patients to radiation for the purpose of medical diagnosis or therapy, or to voluntary participation in medical research programs. Can be exceeded by application to the regulatory agency.

---

10 25 TAC §289.202(n)

11 In Texas, the radiation machine limit is higher. Before the last change in this limit to a lower one, the "general" limit had been 0.5 rem in a year for the TEDE. Many x-ray and accelerator facilities were designed and constructed to comply with that limit. When the limit was lowered to 0.1 rem in a year, the old limit was retained for these facilities to eliminate the need for costly retrofitting or reconstruction.

A member of the general public can only receive a maximum of 100 millirem in a year from a licensee or registrant's radiation operations. For most operations, there is rarely a threat of this level of exposure.

General Effects from External Exposures

Short Term, Long Term Effects

While the effect of radiation exposure on various materials is interesting, we are most concerned with the effects of exposures of humans. There are two effects to be considered when humans are exposed to radiation—short term and long term.

Short term effects involve high radiation doses delivered in short time periods. The results of such doses can be seen ranging from immediately to days or months. The effects range from skin burns/changes to severe body trauma resulting in death. Fortunately, the risk of this type of exposure to the operators/users of most radiation sources is extremely remote and the occurrences are very rare.

Long term effects can develop from either the above type of exposures or from low exposures delivered over long time periods. They may range from cellular damage causing deficient functioning or failure of an organ to cancer. This risk is, perhaps, less remote—but the risk can be easily minimized by proper use of radiation sources and safety equipment and by following reasonable safety precautions.

Biological Effects

When reviewing the potential effects of exposure to radiation, we must remember that the "worst case" circumstances are addressed. The levels of exposure discussed here rarely occur.

Cell Damage

Just like any other material, living cells exposed to ionizing radiation experience ionizations within the molecules of which they are composed. The ions formed react chemically with various components of the cell causing *cell damage*. Generally, low doses of radiation allow the cell to repair the damage quickly. However, higher doses produce more ions and thus, more damage. As the dose increases, eventually, the cell is unable to properly repair the damage – causing cell death or permanent changes. The body can easily replace lost cells, but should a large number of the cells of an organ be severely damaged, then the organ has a higher risk of failure. On the other hand, if the cell's DNA is permanently changed and the cell survives, in some cases cancer may result. Doses of about 50,000 to 100,000 millirem can lead to serious damage to an organ without necessarily causing loss of the organ.

Radiation Syndrome

For doses over 100 rem, "*radiation syndrome* or sickness" may occur. At this level, the gastrointestinal tract is at risk. As the cells of the intestinal lining become damaged, they cannot assimilate food and water. They lose their ability to protect against infection. The exposed person may experience nausea and diarrhea. Over 300 rem, the immune system begins to fail. Over 400 rem, probably 50% of the exposed will die within 60 days if there is no medical treatment. For doses over 1000 rem, it is unlikely that an exposed person will live.

Dose Delivery

The method of dose delivery to the exposed tissue is a major consideration in the effects that the total dose causes. A **chronic dose** is one delivered over a long time period at a low rate or is delivered in small increments over a long time period. An **acute dose** is one that is delivered in a short time period. For example, a person present in a radiation field of 50 mr/hr (0.5 mSv/hr) for 40 hours per week, 50 weeks per year, will receive a chronic total dose of 100,000 mrem (1 Sv) for the year. The person should not experience the symptoms of radiation syndrome listed above—at least not to the degree described. However, if the dose is received in an hour to a day (acute), then many of the symptoms of radiation sickness for a 100 rem (1 Sv) dose should be observed.

Table 2-11 Summary of Radiation Effects

Dose-rem	Dose-Sieverts	Effect
		Low Doses
< 100 rem	< 1 Sv	• usually cell repair • long term effects possible
		High Doses-Radiation Syndrome Or Sickness
> 100 rem	> 1 Sv	• damage to intestinal lining—cannot intake water and nutrients and protect the body against infection • nausea, diarrhea and general weakness

Table 2-11 Summary of Radiation Effects			
Very High Doses-Radiation Syndrome Or Sickness			
> 300 rem	> 3 Sv	•	immune system is damaged and cannot fight off infection and disease
> 400 rem	> 4 Sv	•	about 50% of the exposed are expected to die within 60 days of the exposure if no medical attention is given—due mostly to infection
> 1000 rem	> 10 Sv	•	vascular damage of vital blood providing systems to the nervous tissue (such as the brain)—death probable for 100% of the persons exposed
Note: Doses greater than 100 rem are rare—so risk estimates are based on long term effects from lower doses			

Risk

Whenever we undertake a task, we compare the benefit and the risk—whether we realize it or not. For example, if we need an item from the store and the roads are icy, we review the circumstances to see if the benefit of getting the item is worth taking the risk of driving on icy streets. If we need something frivolous, we probably will choose not to take the risk. However, if we need something essential, such as medicine for a sick child, we are far more likely to take the risk. This same process is used in radiation work. Is the benefit of using radiation and receiving a given level of exposure worth the risk of the potential biological effect?

Risk Assessment

Risk to humans from radiation exposure is estimated by national and international committees using data from evaluations of radiation exposures due to accidents, medical treatments, atomic bombs (Japan), and through human and animal experiments and research. One scientific committee that has made major contributions in risk assessment is the Biological Effects of Ionizing Radiation Committee (BEIR). The findings of this committee, published as BEIR reports, are used by governmental bodies to establish limits of exposure for workers and members of the public.

The problem with risk estimates is they are difficult to make and could be far from the truth. Traditionally, the data used has been from high doses and high dose rates. Further, the doses in many studies had to be estimated, and the studies cannot easily address ethnicity, diet, natural disease, etc.

Estimated Risk
Generally, the risk estimate is for numbers of cancer deaths that are expected to occur due to radiation exposure—not for immediate effects (short term). The BEIR V[12] (report) indicates:

- 0.08% cancer deaths per rem of exposure are expected for acute exposures; and
- 25%-50% less for chronic exposures than the risk for acute exposures

The estimates are for all cancers and ages and include males and females.

Comparison of Risk
The estimated natural death rate of cancer in the US is about 200,000 per 1,000,000 persons. The above indicates that we can expect about 800 cancer deaths to eventually occur in a group of 1,000,000 persons receiving a 1 rem acute dose, or perhaps 200-400 for a chronic dose. So, for a group of 1,000,000 persons to be exposed to a 1 rem acute dose, we would expect a total of 800 cancer deaths.

Radiation Worker Potential Exposures and Risks
Any person working with radiation is naturally going to be concerned with the possible effects that they might experience as a result of exposures received during work. While a few types of radiation duties have a potential for causing exposures of concern, the majority of jobs have a very low risk of causing injurious effects. Users/ operators of industrial radiographic equipment, medical therapy units, certain well-logging radiation sources, accelerators, and workers in large irradiator facilities, face the possibility of receiving high exposures that can cause immediate effects and present serious consequences. Most other operations present a low risk of serious risks.

---

12 Health Effects of Exposure to Low Levels of Ionizing Radiation: BEIR V. *Committee on the Biological Effects of Ionizing Radiation (BEIR V), National Research Council,* The National Academies Press, 1990: P. 6.

On the Job Risks

We are all exposed to natural radiation from our environment (termed background radiation). Radioactive isotopes are present in the soils, vegetation, and all other materials (even our bodies) in the environment. We are continuously bombarded by cosmic radiation. In the U.S., each person receives an annual dose from background radiation ranging from about 100 millirem to nearly 500 millirem (1 mSv to 5 mSv) from these sources. On the average, it is about 350 millirem (350 mSv). In Texas, the annual dose is about 100-150 millirem (1 mSv-1.5 mSv). Most radiation workers, except those listed above, will probably receive less than 150 millirem for an annual dose (TEDE). This is less than the exposure from background radiation.

Example of Worker Risk

How does the exposure from background radiation compare to that of a radiation worker? We can use the case of a cabinet x-ray device operator for comparative purposes. Federal regulations (21 CFR 1020.40) require an emission limit of "0.5 milliroentgens in one hour at any point five centimeters outside the external surface" of a cabinet x-ray system. If a unit operated at this limit (they are usually operated well below this level), then an operator working 8 hours per day, 5 days per week, 50 weeks per year, would receive 0.5 X 8 X 5 X 50 milliroentgens per year, or 1000 milliroentgens, IF he/she was always positioned 5 centimeters from the surface of the unit where this radiation level was present. (Note that 1000 milliroentgens is equivalent to 1000 millirems (10 mSv) for the x-ray energies of the cabinet x-ray units.) Radiation surveys and monitoring find, however, that radiation exposure of operators is much lower than this when the cabinet x-ray unit is properly operated with all safety devices in place and all safety procedures are followed. In fact, a full-time operator should not expect to be exposed to more than 10 millirem (0.1 mSv) per month, or 120 millirem (1.2 mSv) per year. The majority of other users of radiation sources generally experience exposures of less than 500 millirem per year. Industrial radiography and certain medical applications may cause higher exposures.

Compare these exposure levels to the "worst case". A "whole body" dose of 300,000 millirems to an individual in a short time period might cause death within 30-60 days. This is 60 times the annual permissible limit for radiation workers of 5,000 millirems and over 600 times the maximum anticipated annual exposure of 500 millirem (See following table).

Exposure Level	**Approximate Dose**	
	millirem	**mSv**
Life-threatening dose (short term dose)	300,000	3000
Radiation Worker Annual Limit	5,000	50
Annual background dose in Texas	150	1.5
Anticipated maximum annual exposure for full-time cabinet x-ray operator	120	1.2

Table 2-12 Comparison of Doses

As shown in the table above, one can conclude that operating cabinet x-ray units poses little human risk. However, it is important to ensure that the radiation related equipment is in proper operating condition and the operator follows all safe operating procedures.

Chapter 3

Fundamentals of Radiation Protection

1. Introduction

This chapter introduces sources of external and internal exposures and discusses basic tools and radiation safety practices for controlling such exposures.

2. Controlling Radiation Exposures from External Sources

Sources of External Exposures

Most external exposures are caused by discrete sources of radioactive material or by radiation producing machines. To a lesser extent, external exposures occur during work in areas where radioactive materials are manufactured or stored and/or in areas were radioactive waste is stored or processed.

Protective Tools for Controlling External Exposures

Time, Distance, Shielding

When working with radioactive materials (as discrete sources) and radiation producing devices, such as x-ray machines, there are certain steps that can be taken to protect the worker from unnecessary radiation exposure.

In controlling an individual's exposure to a radiation source, three principle protective "tools" or controls are the use of *time*, *distance* and *shielding*.

The amount of **time** that a person is present in a given radiation field determines the total exposure. If an individual is present at a point in an area with a radiation level of 100 mr/hr for two hours, the individual will receive a total exposure of 200 mr. To limit that exposure to only 100 mr, then the individual can be limited to remain at that position for no more than one hour. It is a simple process.

Recalling the Inverse Square Law of Chapter 2, the radiation level drops at points farther and farther from the source. Thus, we can reduce our exposure by increasing the **distance** from the source. If we need to be present in a radiation area for work, in order to minimize our exposure we can maintain the maximum possible distance from the source or we can move the source farther away from the area of work.

The radiation level from any source can be reduced by the use of appropriate **shielding** material. If a work area has unacceptably high radiation levels, shielding material can be instituted or increased to reduce the radiation level to a rate that will not permit excessive exposures to workers or to members of the general public.

These three simple tools can be used individually or in combination to reduce radiation exposure to workers and to keep radiation levels in unrestricted areas to within regulatory limits.

Measuring Exposures from External Sources

Exposures from external sources, where the source is outside of the body and is producing a radiation field which can expose the body, are relatively easy to control by using the time/distance/shielding principles. However, the exposure of the individual must be determined by measurement. Several types of measuring and monitoring devices are available.

Personnel Monitoring Devices

Over the long term, personnel monitoring devices can be used to measure the amount of exposure of an individual to assure that the dose(s) do not exceed any regulatory limit. Three basic devices/techniques used are: film badges, thermoluminescent dosimeters, and pocket dosimeters. Relatively new on the scene is the optically stimulated luminescent dosimeter (OSL), and the electronic dosimeter.

Types of Personnel Monitoring Devices

Film Badges, Thermoluminescent Dosimeters, Pocket Dosimeters, OSLs

Film badges are small packets of photographic film that are constructed and packaged in such a way as to only respond to exposure to gamma and x-rays. They are usually placed in a film badge holder which contains filters for evaluating the type and energy of radiation to account for the RBE—as shown in Figure 3-1. Exposure to gamma and x-ray radiation will cause the film to go through the same physical/chemical changes as exposure to visible light. The effect can be enhanced through special techniques. When a worker may be exposed to significant radiation levels from external sources, the worker will be assigned a film badge. The badge is worn by the worker for a given period of time (usually only a month for film badges). The badge is then sent to the supplier for processing where the film is developed and compared to a known standard (using comparative densities) and the exposure to the badge is determined. This exposure, then, is assumed to be the exposure to the individual that wore the badge for the time period. The advantages of using film badges are that film can become a permanent record of the exposure and the processing equipment can be relatively inexpensive. The disadvantages are their sensitivity to heat[13] and chemicals and "fading" over time. Also, the results are not available until the film badge is processed by the supplier. The use of film badges has declined significantly over the years as new, better methods have been developed.

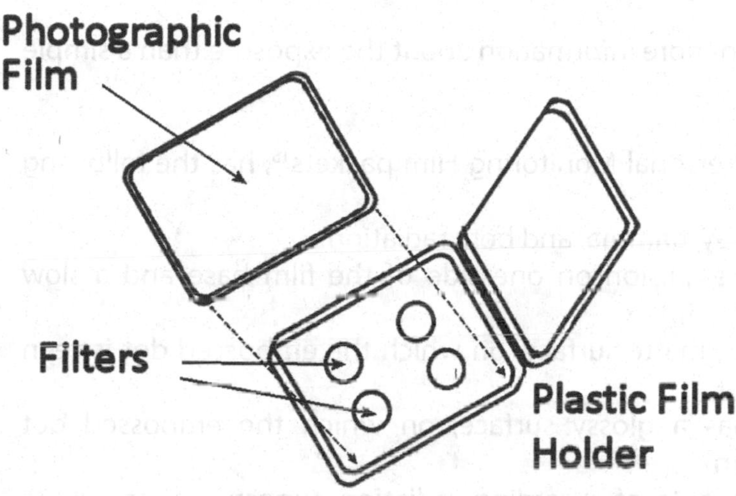

Figure 3-1 Film Badge Holder and Film

A film badge holder[14] supplied by TIMSTAR (RA105600) has six filters. The company describes the function of the filters as:

- An open window which allows all incident radiation that can penetrate the film wrapping to interact with the film. A thin plastic film which attenuates beta radiation but passes all other radiations
- A thick plastic filter which passes all but the lowest energy photon radiation and absorbs all but the highest beta radiation.

---

13 Film badges left on vehicle dash boards have led to apparent, but false, high exposures being reported, causing licensee, registrant, and agency incident investigations being conducted—thereby wasting time and money.

14 TIMSTAR Laboratory Suppliers LTD, Marshfield Bank, Crewe, Cheshire, United Kingdom.

- A dural filter which progressively absorbs photon radiation at energies below 65keV as well as beta radiation.
- A tin/lead filter of a thickness which allows an energy independent dose response of the film over the photon energy range 75keV to 2Mev.
- A cadmium lead filter where the capture of neutrons by cadmium produces gamma rays which blacken the film thus enabling assessment of exposure to neutrons.

Thus the filters help provide much more information about the exposure than a simple film by itself.

A typical film, the Kodak Type 2 Personal Monitoring Film packets[15], has the following capabilities:

- Designed for recording x-ray, gamma, and beta radiation.
- A single film with a fast emulsion on one side of the film base and a slow emulsion on the other:
    - o the fast emulsion has a matte surface, on which the embossed dot is seen as a protrusion;
    - o the slow emulsion has a glossy surface, on which the embossed dot appears as a depression.
- A single piece of film capable of recording radiation exposure over a wide range from less than 30 milliroentgens to approximately 2500 roentgens of high-energy x- or gamma rays.
- Packaging is a paper packet with inner folders surrounding the single film.

The *thermoluminescent dosimeter* (TLD) is another device used to monitor the radiation exposure of the wearer to external sources of radiation. The term thermoluminescent is derived from "thermo", meaning heat, and "luminescent", meaning light. A suitable crystalline material (usually a chip, rectangular in shape and usually a couple of millimeters on each side) exposed to radiation will have electrons of the component atoms change to higher energy states. Later, when the crystalline material is subjected to sufficiently high

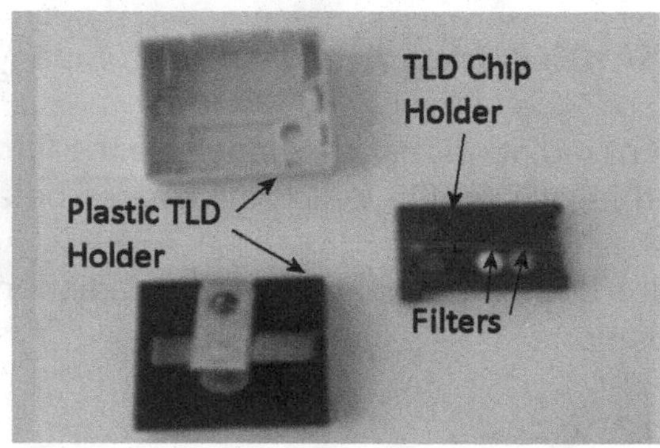

Figure 3- 2 Landauer TLD Holder and TLD

[15] Technical Information Data Sheet: Kodak Personal Monitoring Film, Type 2; TI1480 Reissued 7-04; Eastman Kodak Company, 2004.

temperatures (heat), the electrons will be made to drop back to lower energy states—thereby causing measurable light (photons) to be emitted. The light emitted during the heating process is proportional to the amount of radiation exposure of the material. By comparing the measured amount of light with a known standard, the radiation exposure to the TLD can be determined. This allows determination of the dose to the wearer of the badge that contained the TLD chip (see Figure 3-2). The holders may contain filters to evaluate the energies of radiation observed, in a manner similar to the film badges. The advantages of using TLDs include: unaffected by heat and chemicals; accurate; do not experience fading over time; and easy to use and evaluate results. The disadvantage is that they are a "one-shot" readout. If a problem occurs during the heating process, the exposure information is lost. This problem can be reduced, however, by using several TLDs in the badge. Like the film badge, the results are not available until the TLD is processed by the supplier.

The more recent innovation of the *optically stimulated luminescent* (**OSL**) dosimeter provides for the best features of film badges and TLDs combined. OSLs have the sensitivity, accuracy, and durability of TLDs, and, like film badges, provide for a permanent record as they can be reread. However, the method of operation of the OSL dosimeters will not be discussed here except to note that the system used by Landauer®, a major dosimetry service company, uses a thin layer of aluminum oxide which, after exposure to radiation, is then stimulated by a laser light. The laser light stimulation causes the aluminum oxide to luminance in proportion to the amount of radiation that it was exposed to. The light can be measured and a record made of the exposure.

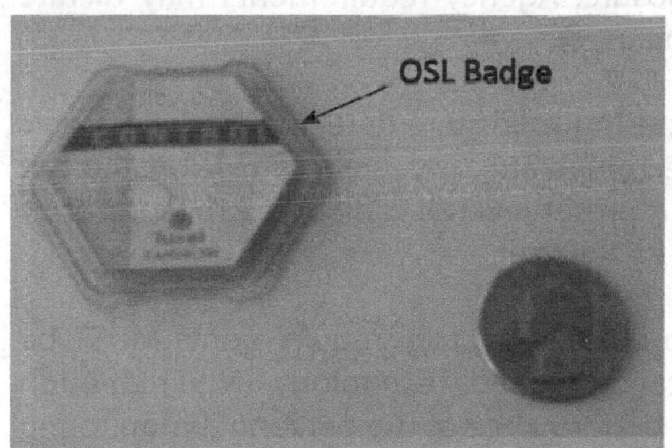

Figure 3-3 Landauer OSL Holder

The most common form of *pocket dosimeter* has been the small ionization chamber constructed as a small, metallic cylinder—about the size of a large writing pen. The end of the dosimeter has a scaled readout. A charging device (see Figure 3-4) is used to set the reading to 0 (called "zeroing") at the beginning of the use or work period. As the dosimeter is exposed to radiation, the internal charges are neutralized by the ionizations occurring in the air of the chamber and the indicator needle changes accordingly. The exposure to the wearer can be determined at any time by holding the opposite end of the dosimeter towards a light and reading the scale. Pocket dosimeters are useful for immediate determination of the worker's exposure and so they are commonly used, even required, for workers with a potential for receiving serious exposures. They are generally used in addition to, not as a substitute for, film badges or TLDs. Disadvantages of pocket dosimeters include difficulty in zeroing, loss of information due to leakage of charge or physical damage, and lack of an independently determined permanent record (the results are often required to be recorded by the worker—but these can be fabricated or altered). More recently, electronic measuring devices have been developed that can provide accurate measurements, audible warnings (of high radiation fields), and a record of exposure. Agency requirements may dictate what type of pocket dosimeter device may be used.

Figure 3-4 Pocket Dosimeters and Charger

Other devices, such as *electronic dosimeters*, using solid state detectors and electronic systems, are also available. They can provide immediate exposure information and can be set to provide an audio/visual alarm at a preset level of radiation exposure. They may be processed, and the results recorded, by the user.

It should be noted that dosimetry used to meet regulatory requirements MUST be accredited by the National Institute of Standards and Technology (NIST) through NVLAP - a stringent system of assuring that performance of the personnel monitoring system meets minimum requirements. However, it is unlikely that any company will offer a personnel monitoring device unless it meets these requirements.

3. *Radiation Survey Instruments*

Detection/Measurement of Radiation and Radioactivity

Radiation surveys are a major tool for protecting persons from receiving unnecessary, even harmful, exposure to radiation. Regulations generally require surveys for key locations and activities that present a potential for causing exposure, contamination, or release of radioactive material into the air, water, and/or soil. There are: (1) surveys to detect or measure radiation levels, and (2) surveys to detect or measure radioactivity.

Detection vs Measurement

Radiation survey instruments have two basic components: a detecting component that somehow responds to ionizing radiation and experiences a change, and an indicator component that displays the change to indicate that radiation was detected. The two components can be integrated together (such as a pocket dosimeter), they can be electrically and/or mechanically connected together (such as a typical radiation survey instrument), or they can be operated separately between detection/ measurement and the indication of radiation (such as film badges or TLDs). If we calibrate an instrument and its detection component using a "known radiation standard", we can use it to measure radiation. Uncalibrated use is considered to be detection of radiation. Without calibration of an instrument, we cannot reliably and accurately measure radiation. We can only determine that radiation is present—or detect it.

Types of Detection

We can detect radiation by a number of techniques, but for day-to-day operations, the portable radiation survey instrument is most convenient. Many such instruments are designed to have interchangeable probes (the detecting component) allowing convenient and economical surveys of multiple types of radiation. Two basic methods used for detection and measurement are *gas ionization* and *scintillation*. There are also solid state methods, but they will not be dealt with here.

Gas Ionization

Gas ionization detectors use a closed chamber with two electric "poles" within. One pole is usually a conducting rod in the center of the (usually) cylindrical detector and the other is the inner walls of the cylinder. The chamber is filled with gas—usually air allowed to pass in and out through a small port, or a specially composed gas that reacts in a desired manner to radiation. The two poles have opposite voltages applied to them so that ions created within the gas by ionizing radiation are attracted to them, creating a "current flow". The resulting current flow

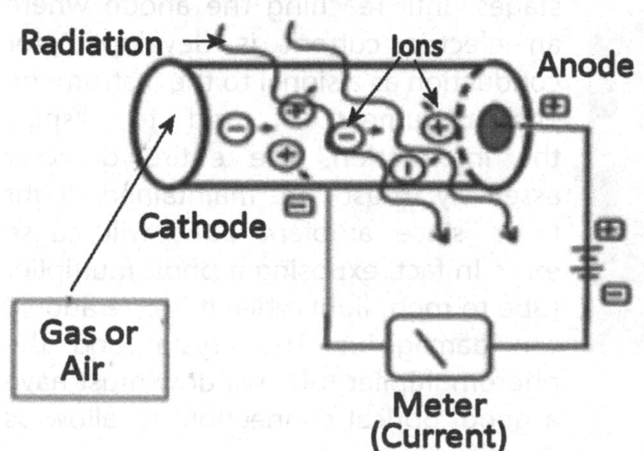

Figure 3-5 Gas Ionization Detector Instrument

changes as the incident radiation increases or decreases and allows the instrument to indicate the detection of radiation. "Ionization chamber" and Geiger-Mueller (GM) instruments use this type of detection.

Scintillation

Some detectors do not use ionization in gases for detection. Scintillation detectors are composed of two parts: a scintillation crystal and photomultiplier tube. The crystal used for gamma radiation is most commonly a solid matrix of sodium-iodide (NaI). The crystalline material will emit light photons when exposed to radiation. The number of photons caused by a given gamma will be dependent on the gammas energy (this effect can be used to identify the distribution of energies of the radiation). The light emitted in the crystal is difficult to detect so it is attached to a photo-multiplier tube (PM tube), using a "light couple", for amplification. This device is capable of using the photoelectric effect to convert light to electrons. This is accomplished by causing the light emitted from the crystal (see Figure 3-7, left object) to enter the PM tube window and strike a photocathode in the "first stage" of the of the PM tube (Figure 3-7, right object). The emitted electrons are then directed down the tube through "cascaded" amplifying stages until reaching the anode where an electric current is developed for conduction as a signal to the instrument. The instrument is used to display the information. The entire detector assembly must be maintained "light tight" since ambient light will cause error. In fact, exposing a photomultiplier tube to room light while it is operational can damage it. The crystal and the photomultiplier tube window must have a good "optical connection" to allow as many photons as possible through to the photocathode.

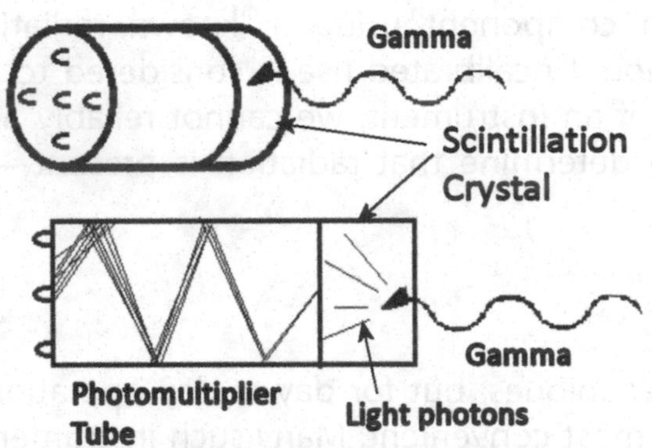

Figure 3-6 Scintillation Crystal and PM Tube

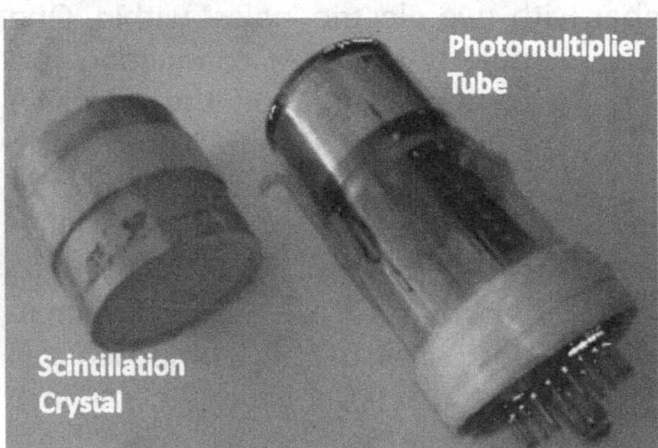

Figure 3-7 Scintillation Detector

One useful tool for identifying various isotopes by the energies of their emissions is the "single-channel analyzer". This instrument will use the amplitude of the signal from the scintillation detector to categorize the energy of the alpha, beta, or gamma that it represents and store the information in a "channel". The analyzer is set to ignore pulses below the low voltage (threshold) setting and those above the high voltage setting so that only pulses representing a specific energy range are counted—the energies in the set channel. The effects of background radiation can be minimized. After a counting time period, the number of counts can then be displayed and compared to the counts of a known source that has been processed the same way—calibration. This allows for identification and quantification of unknown isotopes.

An even more useful variation is to create many "channels" to allow the routing and counting of the signals in a more finely divided array of voltage settings. For example, channel 1 might be set for 0.1 to 0.2 millivolts, channel 2 for 0.2 to 0.3 millivolts, and so on. Hundreds or thousands of such channels can be established with today's electronics. This device is commonly called a "multi-channel analyzer" (MCA). MCAs were once limited to laboratories due to the their fragile construction. Today, they are made to operate with computer components and are not only readily used in the field, but allow for computerized analysis of the data.

Figure 3-8 Ludlum Single Channel Analyzer. —Courtesy of Ludlum Instruments, Inc

Alpha and beta radiation can also be detected through scintillation techniques. "Liquid scintillation" is a common application for detecting and measuring alpha, beta and gamma radiation, but it is used primarily in a laboratory setting. Certain chemicals in liquid form provide for light emission through scintillation. By placing the substance to be evaluated in solution (called a "cocktail"), the resulting photons emitted can be detected by a photomultiplier tube and measured. The system is usually automated and computer controlled. A series of samples can be obtained, placed in cocktails sealed in (usually) 25 ml vials, placed in a conveyor mechanism, processed and analyzed automatically overnight, and the results made readily available to the operator within a few hours.

Survey Instrumentation

Ionization chamber, Geiger-Mueller, Scintillation, Neutron

The choice of instruments for performing surveys depends on a number of factors. The type of radiation, the energies of the radiation, the physical conditions of the survey, the form of the radiation source, and the intent of the survey all affect the process of selecting the appropriate instrument. The predominant radiation survey instruments in use today use the ionization chamber, Geiger-Mueller, and scintillation detection methods of detection, or variations of them. There are others coming into greater and greater use, such as solid state detectors, but, since they are "basic", these three will be only ones discussed here.

The *ionization chamber* instrument (sometimes called a "cutie pie") has a linear response over much of the x-ray and gamma energy spectrum. By linear, we mean that the measured radiation level does not vary as the energy of the incident radiation varies. The ionization chamber of the instrument is not sealed and is subject to pressure and humidity changes as interferences. Normally, the air connection is made by allowing air flow through a desiccator chamber to dry the air and keep the humidity at a minimum—so saturation of the desiccant can introduce operational problems. The instrument has a relatively slow response to detected radiation and radiation level changes. If the instrument's response is full-scale over 6 seconds, measuring an x-ray field of a 1 second duration will show a low, inaccurate, reading. Historically, the chamber of the instrument has been made of low-density material—making it generally sensitive to mechanical shock and not very rugged. Some newer instruments are more rugged and use computer technology to provide both radiation level and integrated dose information.

The *Geiger-Mueller (GM)* instrument uses an ionization chamber that is pressurized with a gas selected for properties that provide a maximum number of ionizations per unit volume. Its response to various energies is not as linear as the ionization chamber, particularly below about 130 KeV (the ionization chamber instrument should be selected over the GM instrument when measuring radiation levels of x-ray devices and gamma sources with energies below 130 KeV). The GM instrument is usually very rugged and has a fairly rapid response to radiation. Full-scale deflection is on the order of about 1 or 2 seconds. The GM instrument is fairly easy to calibrate and, for beta/gamma radiation, can be used for measuring radiation levels. The gas chamber (a metallic cylinder of specific design and material) is usually fixed within the instrument cabinet and/or is fixed within an outer metallic cylinder with a "window" capability. The open window allows measurement of beta and gamma radiation and the closed window limits measurement to primarily gamma radiation. Comparing open and closed window measurements can sometimes give the operator information for identifying the radiation source and assessing the hazard.

Scintillation instruments, being very sensitive, are often used to locate radioactive material in uncontrolled circumstances. Small lost sources, sources at a distance, contaminated surfaces, and contaminated soil often call for the use of a scintillation method. Quantification of the material cannot be accomplished, however, unless the scintillation instrument is calibrated for the material in question. Scintillation probes are delicate, having glass structures internally, and must be handled with great care.

Neutron survey instruments are available, but since their operation is dependent on neutron energies they are difficult to use and their accuracy is often questionable. They are not used as extensively as the above instruments so they will not be discussed here.

Portable Instruments and Probes

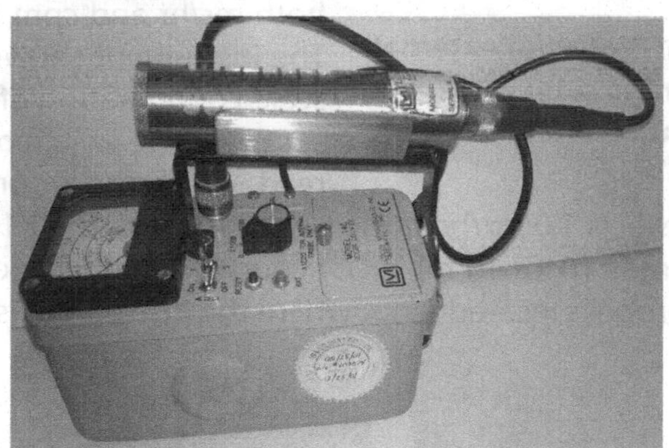

Figure 3-9 Typical Survey Instrument (used with internal or attached external probes). —Courtesy of Ludlum Inst., Inc.

Uses, Operation, Calibration

The portable survey instrument is a major tool in radiation safety programs—although not all programs need them. They are rather inexpensive when compared to other equipment used by companies in their processes of using radiation sources. A typical portable GM survey instrument might cost about $400 to $1,000. Scintillation and ion chamber instruments are generally higher in cost. A basic survey instrument, as shown in Figure 3-9, can have a scintillation or other detector added at relatively low cost (see Figures 3-10, 11). Most ionization chamber instruments have the chamber as an integral component of the unit.

When *operating* an instrument (using the one in Figure 3-9 as an example), we use a shielded cable to connect the GM probe to the electronics of the cabinet. The cable has a "quick connect" on each end and is sufficiently long enough (about 3-4 feet) to allow manipulating the probe freely. Since radiation levels vary greatly, the instrument is designed to indicate radiation levels in ranges—allowing the use of one electronically driven meter (or digital readout). A switch is used to select a range of measurement or scale. If the switch is set too low, the meter will indicate a maximum (full-scale) reading. Set too high, the radiation level will be indicated as zero. Typical switch markings indicate X1, X10, . . . X100 (or more) ranges. The meter face may have one or several scales on it. It may also be scaled for readings of both mr/hr and cpm (counts per minute). The instrument shown in Figure 3-5 has a scale configuration of 0 to 2.0 mr/hr. If we set the range switch on the X1 setting and read 1.0 mr/hr on the meter, the measured radiation level is 1.0 mr/hr. If the switch is set to the X10 range, the reading would be 1.0 mr/hr times 10, or 10 mr/hr. Of course, the indicated radiation level is at the point where the probe is located—not where the instrument is positioned.

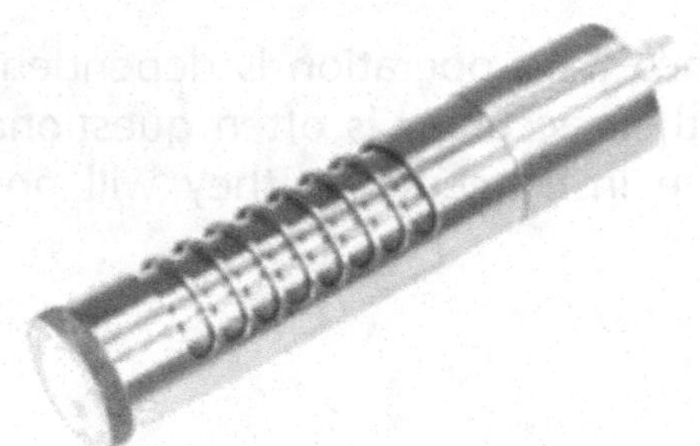

Figure 3-10 Typical GM Probe (for external attachment to survey instrument).

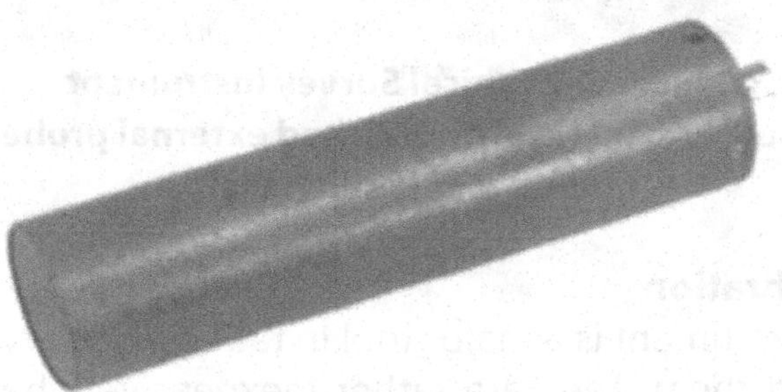

Figure 3-11 Scintillation Probe (for attachment to instrument). —Courtesy of Ludlum Instruments, Inc

The following are examples of survey instrument readings:

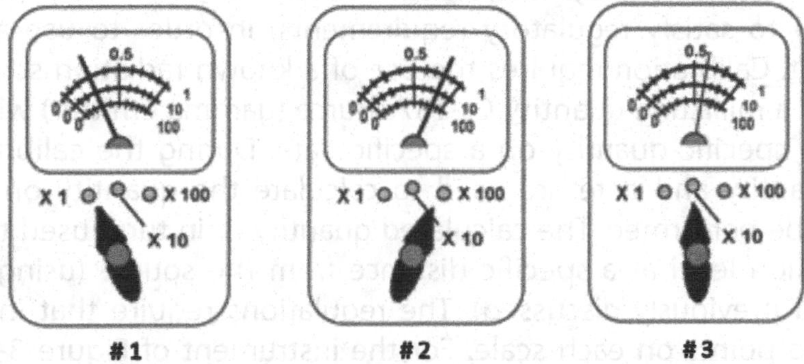

Figure 3-12 Survey Instrument Example Readings

- Instrument #1 has the switch set to X 1, so on the 0-1 scale the needle on the meter reads 3. The reading would be 0.3 units.

- Instrument #2 has the switch set to X 100, so on the 0-100 scale the needle on the meter reads 8. The reading would be 80 units.

- Instrument #3 has the switch set to X 10, so on the 0-10 scale the needle on the meter reads 6. The reading would be 6.0 units.

This is a simplified scale. Often, the scale is chosen according to the type of detector attached. One scale may be for measuring mr/hr (mSv/hr) using a GM detector, while another may be scaled for "cpm" (counts per minute) for scintillators, alpha, and beta detection probes.

If a scintillation probe is attached, such as the one shown in Figure 3-7, the cpm (counts per minute) scale is used. While this combination is used primarily for detection (not measurement in this configuration), the surveyor can learn to make a "rough assessment" of the level of hazard through experience.

The same instrument can accept a beta detector using the same cable. While we can detect betas, if we want to measure or quantify the beta field, we will need to perform a calibration by using a known beta source. Many instruments, today, have digital readouts instead of a meter. These are much easier to read. Some of these can provide a great deal of additional information, such as isotope identification. However, one still must take care to assure that the correct switch settings are selected and the instrument is in current calibration to assure correct results.

Instrument Calibration

Calibration of radiation survey instruments must be performed (by an agency authorized party to satisfy regulatory requirements) in order to use an instrument for measurement. Calibration requires the use of a known radiation source. A typical source might be a millicurie quantity Cs-137 source (gamma emitter) which has been certified to be a specific quantity on a specific date. During the calibration process, the certified quantity and date are used to calculate the quantity on the date the calibration is to be performed. The calculated quantity is, in turn, used to provide the calculated radiation level at a specific distance from the source (using the distance squared method previously discussed). The regulations require that instruments be calibrated at two points on each scale. For the instrument of Figure 3-8, two points that would be selected on the 0-2 scale would be 0.5 and 1.5. The first is about 1/3 across the scale from the 0 position and the second is about 2/3's across. We would then calculate the distance from the calibration source to produce the desired radiation levels.

Figure 3-13 Instrument Calibration Example

The following setup might be used for calibrating the above instrument:

Table 3-1 Example of Instrument Calibration Settings

Switch Setting or Range	Desired Radiation Level (mr/hr) [2 points]	Calculated Distance from 100 mCi Cs-137* (meters)
X0.1	.05	25.69
"	0.15	14.83
X1	0.5	8.12
"	1.5	4.69
X10	5.0	2.57
"	15.0	1.48
(source)	33	1.00
X100	50	0.81
"	150	0.47
X1000	500	0.26
"	1500	0.15

*** .33 R/hr per Ci @ 1 meter or 330 m R/hr per Ci @ 1 meter**

The radiation source would be set a position and a range marked off with the calculated distances being marked for each desired calibration point. The probe of the instrument is placed at each point and the radiation level read. Some instruments can be adjusted to provide the proper indication. If not adjustable, the reading must be recorded and a calibration curve developed so that later readings (during use) can be corrected. For the X1000 scale, it becomes difficult to steady the probe at the small the distances used. A small movement of the detector causes a large change on the readout. The regulations require that the instruments be calibrated to "plus or minus 20 percent of full scale". Full scale for the X10 range is 20 mr/hr—so" 20% would be 4 mr/hr. On adjusting the target low point of 5 mr/hr, a readout of 1 to 9 mr/hr would be acceptable. For the high point of 15 mr/hr, a reading of 11 mr/hr to 19 mr/hr would be acceptable. During the calibration procedure, the person performing the calibration should be wearing his/her personnel monitoring device (film badge or TLD) and should wear a direct readout monitoring device, such as pocket dosimeter. Ionization chamber instruments are calibrated in the same manner, although in some cases a calibrated x-ray device may provide the known radiation fields.

Radiation Survey Instrument Use

GM and ionization chamber radiation survey instruments are usually the instruments of choice for performing surveys to meet regulatory requirements. For industrial radiography and well-logging, the instruments are required to be calibrated at least every six months as they are used in operations involving large and dangerous radiation sources and they are used extensively in the field where they are subjected to rough handling. Most other uses require instrument calibration annually. As a general rule of thumb, instruments used in field operations should be calibrated semi-annually and instruments used in-house should be calibrated annually. If not specifically addressed in the regulations, the calibration frequency will be set in the licensing or registration process. A radiation survey instrument may not be used to satisfy regulatory requirements unless its calibration is current. If a licensee/registrant possesses only one instrument, then the calibration frequency must be diligently followed. Depending on your type of use of radiation, if the instrument is out of calibration for one day, it may be considered a violation—even though the instrument may not have been used. On the other hand, a licensee/registrant having more instruments available than required, can have calibrations "lapse" without being in violation—provided an instrument is not used when not in current calibration.

When using a GM or ionization chamber instrument where the instrument's proper operation is essential to safety, a *check source* may be employed to verify that the instrument is working properly. This IS NOT a substitute for calibration! Check sources only verify that the instrument is detecting radiation at the time of use and the readout device is so indicating. Most check sources, which are usually small plastic discs about the size of a quarter, are "exempt" quantities (do not require a specific radioactive material license).

Figure 3-14 Typical Check Source

As with any tool, the different instruments have their limitations and advantages. The primary limitation of the GM instrument is its non-linear response to x-ray and gamma radiation of energies below 130 KeV, although it is a rugged instrument with a good response time. Limitations for ionization chamber instruments include a slow response time, sensitivity to shock, and air temperature, pressure, and humidity changes, although they do have a very linear response. The limitations for scintillation instruments include sensitivity to mechanical shock and inability to determine quantity without elaborate calibration. However, scintillation instruments are very sensitive and can detect very small quantities of radioactive material.

Performing Radiation Surveys

Survey Techniques, Discrete Sources, Contaminated Surfaces

The techniques selected for performing radiation surveys usually address two general conditions: radiation levels emanating from an x-ray device or discrete source (contained in a small volume—such as a sealed source or a vial of radioactive material), or radiation levels from spots or areas with radioactive materials on them (such as contaminated surfaces). The usual purpose of any survey is to show that radiation levels (or radioactive material concentrations) meet regulatory limits in areas were persons work, or in areas that members of the general public may have access to.

Surveying radiation levels around *discrete sources* is often the easier task. Generally, a calibrated survey instrument is used and the instrument's measurements can be easily read and documented. In some operations, the survey instrument is the primary tool protecting workers from unnecessary, even harmful, radiation exposures. In this case, the instrument should always be readily available and its operation constantly checked. Should the instrument appear to fail or operate improperly, the radiation worker should stop operations, secure the radiation source, and contact the company's radiation safety officer for assistance. In general, key survey points should be pre-selected and surveyed with the highest observed readings being recorded. Setting up appropriate survey points is part of the licensing process. The survey record or document should indicate both the radiation reading and the point at which the reading was obtained (i.e., wall surface identity, distance from the wall surface and height above ground where measurement is taken). Other information generally required on a record would be the identification of the survey instrument used, identification of the person making the record or performing the survey, the date of the survey, and, in some cases, information regarding the radiation source(s).

Surveys of radiation levels performed around or near non-discrete sources usually involve *contaminated surfaces*. The distance-squared relationship only applies to small sources, so measuring general radiation levels over surfaces with varied levels of radioactive contaminants can be nearly meaningless—although sometimes useful for general assessment of a problem area. An additional problem with these surveys is the elevated potential for contamination of the instrument's detector. A contaminated probe will provide a continuously higher result and make it useless. Masking tape, spacers, or other protective material can be placed on the probe at points likely to touch surfaces in order to reduce this potential. For fixed contamination, large surface area detectors can be used to measure the radiation levels or emissions fairly accurately.

As an example, in considering release of a facility or equipment for unrestricted use in which Am-241 had been used, we can verify that alpha fixed contamination limits are not exceeded by using a 5 inch alpha probe to measure the surface contamination—following calibration with an alpha source. The "acceptable[16] surface contamination levels" listed for Am-241, a transuranic, are:

- Average 1,000 dpm/100 cm^2
- Maximum 3,000 dpm/100 cm^2
- Removable 200 dpm/100 cm^2

Since this type of probe has a surface area of 50 cm^2 (cm means centimeter), it is relatively easy to calibrate the instrument and quickly perform a survey at numerous points. An alpha calibration source is used to determine the calibration factor (CF)—or the number of dpm (disintegrations per minute) which results from a given measurement in cpm (counts per minute) by the probe. Then each result can be calculated as:

$$\text{\# dpm per } 100\ cm^2 = 2 \text{ X cpm X CF dpm/cpm}$$

an easy calculation.

For fixed contamination, if the average of all of the results obtained were less than 1,000 dpm per 100 cm^2, and no SINGLE measurement on the surface exceeded 3,000 dpm per 100 cm^2, then the surface of the facility or equipment would surveyed would be considered releasable.

For removable contamination, a different technique is necessary and one might use a 1 inch probe instead of a 5 inch one. The amount of removable material must be determined. This can be accomplished by wiping a 100 cm^2 area and surveying the

16 TAC §289.202(ggg)(6)], October 2011.

wipe. The wipe can be taken using filter paper or commercially available products, such as sticky material of 100 cm^2 area. The filter media (filter paper) is generally placed in a stand device/sample holder which positions the surface of the paper within a centimeter of the probe surface. The same stand should be used to perform the calibration as above and the distances, from the probe to the sample surface, should be held constant for calibration and each measurement. The calculations are performed in the same manner as for the fixed contamination procedure—except multiplication by 2 is not performed because the surface of the wiping media represents the total particulate for 100 cm^2 (in this example).

4. Controlling Radiation Exposures from Internal Sources

Exposures from Internal Sources

For most uses of unsealed forms of radioactive material, it is relatively easy to control internal exposures. We simply use systems and procedures that keep radioactive materials out of the body. However, when radioactive material does get into the body, measurement, evaluation, and analysis can be very complex.

Internal Contaminants: Biological Half-Life, Radiological Half-Life, Effective Half-Life

Radioactive atoms and molecules that enter the body will remain there for a time which is dependent on their chemical and physical forms and their solubility. The time required for half of the material to be removed from the body, without regard for decay, is called the *biological half-life*. We have already discussed the *radiological half-life*. If we take into account the biological and radiological half-lives of a given isotope, after a time period we will find that the total remaining amount of the isotope will be dependent on both. Periodic measurements will show that the total remaining amount decreases over time—but a decrease different than predicted by the biological or radiological half-lives considered separately. This is due to the effective half-life. The *effective half-life* is less than the smaller half-life of the two. For example, if an isotope had a radiological half-life of 1 day and a biological half-life of 1 week, the effective half-life would be less than 1 day.

Isotopes that are alpha emitters are particularly serious as internal contaminants. The large particles give up their energy over a short distance in the tissues as they interact easily and have a high LET. This allows many ions to be formed in small volumes, such as a single cell. It is difficult for individual cells to repair the resulting damage. Some isotopes will be concentrated into specific organs more so than in other organs. For example, the thyroid gland concentrates iodine more than any other organ, so radioactive iodines will be concentrated there. Internal Beta (causing Bremsstrahlung radiation) and gamma emitters also present a "whole body" exposure problem in addition to the localized exposure.

Pathways of Internal Exposures

Inhalation, Ingestion, Skin Absorption, Open Wounds, Punctures

In both preventing internal exposure and in evaluating an exposure that may have occurred, the pathway (the method of introducing the radioactive material into the body) is a major consideration. Radioactive material can enter the body by inhalation, ingestion, absorption through the skin, injection, or through an open wound.

Any airborne radioactive contaminant can be a potential hazard by *inhalation* through the nose or mouth. Where and when the contaminant accumulates within the body depends primarily on particle size and chemical form. Larger particles collect in the upper respiratory passageways and are usually transported to the digestive tract (see Ingestion pathway). Smaller inhaled particles can be deposited in the lower tract (bronchial tubes and lung tissue) causing exposure to the tissues that they become adhered to. Eventually, the body is able to process most of these particles out of the tissue using the lymphatic system, although it is usually a very long time period and subjects the surrounding tissues to prolonged exposure. Alpha emitting radioactive materials are a serious problem for this pathway.

Most *ingestion* of radioactive material probably occurs when a person places a contaminated item or a contaminated hand in or around the mouth. Pens, pencils, cigarettes, and pipes that become contaminated in work areas or from a worker's hands are common mechanisms. Also, larger particles of inhaled contaminants can be introduced to the digestive tract. Once a contaminant is in the digestive tract, the chemical form and solubility determine whether all, most, or none will be absorbed or will pass on through and out of the body. These also are factors in the length of time that the contaminant remains within the body. Once in the digestive tract, if the chemical form of the isotope is soluble, it may be absorbed into the body along with the nutrients processed in the tract. Insoluble forms may pass on through causing only short term exposure to the nearby tissues. Once absorbed or processed, the different organs' individual preferences will determine the concentrations that will be observed. A radioactive contaminant concentrated in an organ can then be a source of exposure to the organ itself, to nearby organs, and the entire body. Occasionally, other persons can also be exposed.

Some isotopes can be readily *absorbed through the skin*. Once skin absorption occurs, they can be distributed by the circulatory system and become concentrated according to organ preference until they are eliminated.

Radioactive material is intentionally *injected* into the body during certain medical procedures. Otherwise, the only other method of getting radioisotopes past the skin barrier and into the body is through an accidental break in the skin. An existing *open wound* or *puncture* by a sharp, contaminated object may provide an avenue. The body may then process the material as it does any other chemically similar material or it may remove the contaminant as an unwanted material.

Determining Doses from Internal Sources

It is quite difficult to determine the dose to the body and individual organs when internal contaminants are known or suspected. Both in vivo and in vitro techniques are available to measure the amount of contaminant within the body. Once the quantity of contaminant is determined, the results must be analyzed to find the dose(s) delivered to the body and/or organ(s). In most cases, assistance from a qualified physician (radiologist, for example), Certified Health Physicist (CHP) or Licensed Medical Physicist (LMP) will be needed.

Dose Measurement

Depending on the isotope and its form, both non-invasive and invasive techniques can be used to measure the quantity of contaminant taken into the body.

Bioassays are performed by obtaining samples of body/organ tissues or of body fluids. In most cases, non-invasive procedures are used. Urine, saliva, perspiration, and fecal samples can be obtained and analyzed for radioactive content. On rare occasions, it may be necessary to obtain a tissue sample through a surgical procedure conducted by a physician—much the same as a biopsy is performed.

Some organs have preference for certain radioactive materials (chemical form and solubility must be considered). The thyroid gland has a natural preference for the element iodine and will concentrate more than any other part of the body. If the iodine happens to be radioactive, then the thyroid will become a small "radiation source". A detector can be placed in close proximity to the thyroid gland and the gamma radiation "counted"—thus the term *thyroid count*. This allows determination of the amount at the time of the assay. Several counting sessions may be necessary to obtain data for a more complete assessment.

One method of determining the radioactive content of a person's body is to measure the gamma radiation being emitted from the body. Large, sensitive detectors, or an array of such detectors, can be placed in a well-shielded chamber/facility and the person's body placed within the chamber. The number and energy of gammas being emitted from the body can be measured and the type and quantity of radioisotope determined. *Whole body counters* are expensive to acquire and operate so they are not readily available. They can be found at large radiation use facilities, national research centers, and some nuclear power plants. They are often made available for emergency situations. Some units are designed to count major portions of the body instead of the whole body.

Analysis of Bioassays

After measuring or determining the isotope and quantity of radioactive material that a person has taken into the body or organ, the task is not complete. Recalling the relative half-life, we know that the concentration will not be the same during and after the intake. The intake may occur as a "one-time only" circumstance—or it may occur daily over a long time period. Considering also the chemical and physical forms, the type of intake, organ preference, and time of measurement, we find that there are many complicating factors. In most cases, the doses can only be estimated.

After the dose has been estimated, the *potential biological effect* to the body or organ must be considered. For example, will the dose received by a thyroid exposure to radioiodines be harmless, cause reduced thyroid function, or destroy it totally? This information must be made available to the physician to obtain proper medical treatment. A resulting injury may not be due to the isotope's *radiotoxicity*. For example, in the uranium mining and ore processing industries, urinalysis is the usual bioassay requirement for workers. The action limit for positive results found, however, is based on the *chemical toxicity* of uranium—not its radiotoxicity. This is due to the kidney being damaged more by chemical toxicity than any other tissues by radiotoxicity.

5. Radiation Safety Practices for Controlling Internal Exposures

As previously discussed in Section 1, controlling external doses is rather easy. However, controlling internal doses requires more diligence. The best method of controlling internal exposures is to prevent introduction of any radioisotopes into the body. To accomplish this, contamination must be minimized during work and sealed sources (containing radioactive material) that may be leaking must not be put into use.

Prevention of Contamination from Unsealed Radioactive Material

Use of proper precautions and tools can prevent contamination. When working with radioactive materials that are not sealed, there are many standard safety procedures and techniques that can be followed and tools that can be used for self protection.

Airborne Radioactive Materials (Inhalants), Contaminants

To protect against inhalation of *airborne radioactive materials and contaminants*, work should be performed in a fume hood or glove box or in an area with well controlled ventilation. In most cases, a standard laboratory fume hood, with properly designed exhaust ducting drawing off the potentially contaminated air through a filtering system, will provide sufficient protection. Glove boxes can be outfitted with HEPA and charcoal filter systems to protect the breathing zone. These systems should be designed and tested by a qualified individual. Respiratory protection devices, such as face masks with filters, contained air supplies, etc., may need to be considered. However, the use of these devices will require additional procedures—particularly the implementation of a mask fitting and control procedure. Periodic or continuous sampling of air may need to be performed. The radioactive concentration in air can be controlled.

Surface Radioactive Contamination

To protect against ingestion of radioactive material, one must avoid contaminating work surfaces and areas, as well as the hands, face, clothing, and tools. One must also avoid using equipment that may lead to ingestion. The wearing of laboratory coats or aprons, protective gloves, shoe covers, etc., should eliminate or minimize contamination of hands, skin, and clothing. Even when wearing protective gloves, improper use of equipment can lead to self-contamination. A worker wearing gloves that are contaminated during work with radioactive material can pick up a telephone receiver, use a pen or pencil, turn on/off a light switch, etc., contaminating them in a manner that can later lead to contamination of the worker—or other persons. Any person later touching those items with their unprotected hand will most assuredly become contaminated and the contaminant will probably eventually be ingested. Using the proper tools in the correct manner can prevent or minimize contamination. When performing pipetting procedures, auto-pipettes must be used to avoid accidental ingestion. Generally, pipetting by mouth is prohibited when working with any hazardous material (as with mouth pipetting). Performing work with gloved hands and using auto-pipettes will prevent most ingestion. If a procedure requires the use of remote-handling tools, then those tools must be used.

General Protective Procedures

Often, it is necessary to establish a "hot line" at the boundaries of potentially contaminated areas to control releases. By donning protective equipment (lab coats, gloves, shoe covers, etc.) at the hot-line entrance to the area and then properly removing it at the hot-line exit from the area, the radioactive material can be kept in the controlled area. The potentially contaminated protective equipment should be bagged for laundering and/or disposal. If laundering is performed, then either a specialized licensed laundry must be used or the garments must be stored for decay if only short lived radioactive material is used. Special license procedures and conditions are usually applied.

Prevention of Contamination from Sealed Sources of Radioactive Material

Sources of radiation that are designed to keep the radioisotope contained (sealed or plated) must be tested regularly to assure that their method of containment remains effective and no material is leaking. Leak tests are usually required on 6 months, 1 year, or 3 year intervals as required by the agency—based on the manufacturer's specifications.

Forms of Radioactive Material

Radioactive material occurs or is produced in a number of forms. It may be liquid or in solution; solid or powder; gaseous; or nearly pure or mixed with other elements, isotopes, or molecules. When it is desirable to use ONLY the radiation emitted from a radioisotope (not the radioisotope itself), then the radioisotope can be sealed into a container termed a "sealed source" (see regulatory definition below). In some cases it may be electroplated onto a metallic surface. A sealed source is usually composed of one, two, or even three, stainless steel right cylinders into which a radioisotope has been placed and the end(s) welded closed. Unfortunately, not all such arrangements have perfect seals. A defect in the weld(s) or sealing component may occur during manufacture or the sealed source may be damaged during use under harsh conditions. Build up of pressure, caused by production of gases and/or heat buildup during decay, may force some of the radioisotope out of the sealed source. The resulting contamination can cost millions of dollars to clean up. Occasionally, the surface of a sealed source is contaminated by a different radioisotope during manufacturing. Periodically, sealed sources must be tested for leakage or loss of integrity to prevent possible contamination.

Regulatory Definition of Sealed Source*[17]*: *Sealed source—Radioactive material that is permanently bonded or fixed in a capsule or matrix designed to prevent release and dispersal of the radioactive material.*

[17] TAC §289.201(b)(97) [April 2003]

Leak Tests

A sealed source may be mounted in an operating device or remain in a storage container—both providing shielding while limiting direct access to the source. In some cases, the source can be *leak tested* by direct access. There are a number of variations.

Leak tests are performed from every six (6) months to every 36 months, depending on the specific regulatory requirements. A six (6) month period is the usual requirement.

A general procedure for leak testing is:

- A wipe of the sealed source surface, or the nearest surface of the container or device to the source (such as a beam port), is performed using a cotton swab, Q-tip, or similar material. The wiping media may be moistened with water or alcohol.
- The swab is then placed in a plastic bag, removed from the radiation area, and surveyed to assure that it does not have excessive quantities of radioactive material on it.
- If the survey indicates the radioactive content of the swab doesn't exceed postal regulations, it can be mailed to the service company for analysis (some licensees are authorized to perform their own analysis).

Note: Service companies usually send the customer a kit for each source that is due to be leak tested. The kit usually contains instructions for obtaining the wipe and sending it back to the service company.

If the results of leak testing indicate that the following limits are exceeded, then the regulations require that the source be removed from service until repaired:

- The presence of 0.005 μCi (185 Bq) or more of removable contamination on any test sample.
- Leakage of 0.001 μCi (37 Bq) of radon-222 per 24 hours for brachytherapy sources manufactured to contain radium.
- The presence of removable contamination resulting from the decay of 0.005 μCi (185 Bq) or more of radium.

Methods of Decontamination

There are two categories of contamination: personnel contamination and contamination of equipment and facilities. The contamination can occur in both restricted and unrestricted areas, such as the environment around a facility.

Decontamination of Personnel

Personnel can be contaminated on their skin and hair, or contaminated within their bodies—although we do not normally address the latter as "contamination".

External Contamination

Decontamination of a person's body surface can be readily and successfully performed. If there are no wounds, removal of contaminated clothing and cleansing of the contaminated surfaces with good (non-abrasive) detergents and mildly warm water will usually be satisfactory. Care must be taken to avoid breaking the skin and to prevent inhalation or ingestion of cleaning by-products during the cleaning process. Radiation surveys should be performed periodically to assess the progress of decontamination. Both the contaminated person and personnel participating in the procedure should be properly equipped to prevent re-contamination. If the contaminated person has serious injuries or illness, these conditions should be medically attended to first. Medical personnel should be advised of the contamination problem and provided with appropriate protective equipment. All decontamination materials should be collected and bagged to prevent further spread of contamination.

Internal Contamination

All procedures for removing radioactive material from within a person's body should be performed under the supervision of a licensed physician qualified in handling radiation emergencies. There are only a few facilities within the US that are both qualified and equipped to deal with major radiation medical emergencies. The state (with jurisdiction) radiation control agency or the US Nuclear Regulatory Commission can assist in finding the appropriate medical resources. Licensees with "high risk" operations should identify these resources in advance and incorporate them into their procedures.

For radioactive materials deposited in the digestive tract or the respiratory system, medical procedures can be used to remove a great deal of the contaminant. When radioactive material has been absorbed into the body, chelation may be used to remove some of the contaminant. Chelation is a process of injecting the appropriate media (chemicals or drugs) into the body to "tie up" the radioactive atoms. The chelation media is then eliminated from the body, thereby removing the radioactive material. This procedure has had mixed results. Chelation is a medical procedure and must be performed by qualified medical personnel.

Fluids containing radioactive material, removed from a person, must be collected and handled as radioactive waste. The person's excreta may be disposed through the sanitary sewerage system.

Decontamination of Equipment and Facilities

Generally, limits are imposed by a licensee (in the safety procedures submitted with license application) for radioactive contamination that may be caused during normal operations. The limits apply to the restricted or controlled area—as in a "hot lab". For unrestricted areas (areas not under the licensee's control), regulatory limits apply. Each licensee should have procedures for periodically evaluating both of these types of areas—as well as procedures for decontaminating them when the limits are exceeded. There should also be procedures for decontaminating tools and equipment that may become contaminated either during use or during a radiation accident.

Most contamination problems involve surface contamination—which may be loose or fixed.

Types of Contamination

Loose, Fixed

Loose contamination consists of radioactivity that is easily removed from a surface and can be readily spread to other surfaces through contact. *Fixed contamination* is adhered to the surface and cannot be readily removed by the use of standard decontamination methods.

The objective is to remove as much of the radioactive contamination as possible—NOT spread the contamination.

General Procedures for Decontamination of Loose Contaminants (particularly following a radiation accident)

> *Develop a Plan*—A well constructed plan will allow for a timely recovery of the area at a minimum of exposure to personnel. Discuss it with the regulatory agency.

Provide Basic Protective Equipment—Assure that each decontamination worker is properly monitored with film badge or TLD and appropriate survey instruments are available. Pocket dosimeters may be needed to control exposures.

Provide Proper Survey Equipment—Set appropriate radiation survey and analytical equipment at positions where they will not become contaminated. Treat wipes of surfaces as contaminated.

Protect Survey Equipment—Radiation survey instruments and probes may need plastic covers to prevent contamination.

Control Areas—confine the area where decontamination is to be performed by establishing boundaries (hot-line) and controls (such as warning signs, ropes, tapes, etc.). Walls and doors of a room may be sufficient.

Establish Control Lines—set up control points for entry and exit to the controlled area:

Entrance—point to don protective equipment.

Exit—point to remove potentially contaminated equipment and survey personnel.

Require Protective Clothing—decontamination personnel should wear anti-contamination clothing or lab coats/aprons and protective gloves and shoe covers.

Consider Respiratory Protection—In some cases, respiratory protection may be necessary.

Protect Open Wounds—Workers' open wounds should be covered or bandaged to prevent internal contamination (if they are required to be in the area for some reason).

Label Containers—clearly label all containers of radioactive materials or waste and known contaminated and potentially contaminated items. Label rad waste containers for solids or liquids.

Protect Uncontaminated Surfaces—place plastic or absorbent material over, under and around decontamination work areas and equipment, as necessary, to avoid spread of contamination.

Assure Fume Hoods are Operational—when handling potentially volatile chemical forms of radioisotopes, fume hoods should be used and should be in proper operating condition.

Periodically Monitor Personnel and Areas—Assure that personnel do not become excessively contaminated. If used, check pocket dosimeters frequently.

Hint: Clean from less contaminated areas in towards the more heavily contaminated areas to reduce opportunities for spread of contamination. Also, replace cleaning tools (mops, sponges, etc.) frequently so that they do not become a source of contamination. Use disposable cleaning tools.

General Procedure for Decontamination of Fixed Material

Removal of fixed contamination requires the same procedural steps listed above for loose contamination. Additional steps and precautions are:

Choose Appropriate Cleaning Tools—methods for removing fixed contamination might be wiping, scrubbing, flushing, and/or soaking. DO NOT use grinding, sanding, scraping, or chipping methods without using respiratory protection.

Disassemble Complex Devices—Removal of contaminated parts for individual decontamination should reduce the likelihood of spread of contamination to the entire unit—making it unusable.

Hint: It may be less expensive to replace a piece of equipment than to spend hours or days decontaminating it.

Consider Alternatives—Some contaminants may not be removable to within limits. Protective cover may be all that is needed under certain circumstances—particularly for items that will only be used in restricted areas. For example, contamination of an item by radioisotopes whose only hazard is alpha emissions might be painted as the paint will prevent emission of the alphas. Usually, approval by the regulatory agency is required.

Post-Decontamination Procedure

Place radioactive materials and waste in properly labeled containers and secure them. Secure containers in proper storage facilities.

Survey Personnel—monitor personnel exiting the contaminated areas after they have removed protective equipment to assure that they have not become contaminated.

Complete Surveys—Perform final surveys and surface wipes. Analyze wipes and record results.

Review Results—Review the survey results to assure that decontamination efforts were successful and all areas are now in compliance with regulatory requirements.

Document Activities—Record all activities and surveys and maintain documents.

Notify Agency—Should the contamination appear to be extensive, or if found to be more extensive during decontamination activities, the agency with jurisdiction should be notified. Most agencies can provide assistance.

The objective is to remove as much of the radioactive contamination as possible—NOT spread the contamination.

6. Special Equipment and Monitoring

Protection Equipment

Source Handling Tools

Personnel exposure and contamination can readily be minimized by the use of proper handling and protective equipment.

Remote handling tools rely on the "distance" principle for reducing exposure to the operator. A remote handling tool should provide a positive control or grip on the radiation source, be reasonably easy to manipulate, and should provide sufficient separation between the source and the operator to minimize exposure.

Many sealed sources intended to be removed from the container have a "*designed handling tool*" for manipulating the source. Some may be a rod with a gripping socket on the end which matches the source holder end and providing a foot or two of separation between the surface of the source and the operator's hand. In some cases, a tool with fingers on one end of a flexible shaft and a manipulating control on the other end provide for good separation and control. Tongs with long handles may be the tool of choice for many operations. Radioactive material MUST NOT be handled directly by any person.

Protective Devices

In addition to designed fixed shields, portable shields and shielding aprons can be used in operations to reduce exposures of workers. For temporary or emergency conditions involving gamma emitters, cinder block, sand, bricks, lead sheet, lead shot, and water can be used to provide adequate shielding. Similarly, water or blocks of paraffin may be used for neutron sources. Resources are usually available locally.

Protective Clothes and Equipment

To protect against personnel contamination when using loose radioactive material or potentially contaminated tools and equipment, proper selection of protective equipment is necessary. Anti-contamination clothes, coveralls, aprons, lab coats, protective gloves, and protective shoe covers should be considered when setting up procedures or when recovering from accidents. If airborne hazards are indicated, protective masks may be necessary. Fortunately, licensees that conduct operations where these items may be needed should have already provided for them and will have prepared procedures for proper use when applying for their radioactive material licenses.

Monitoring Equipment and Procedures

Some licensed operations using radioactive material present hazards of both external and internal contamination. Usually, only larger, complex programs have these potentially hazardous processes. Fortunately, they will also have the resources for dealing with these hazards. A brief overview is provided here.

Environmental Monitoring

For releases of radioactive material into the environment, regulatory limits have been established. By controlling emissions of radioactivity into air, water, soil, and sewage systems, exposures from both external and internal sources can be prevented or minimized. However, the effectiveness of such controls can only be measured through proper periodic monitoring procedures—i.e., *environmental monitoring*. Radiation control regulations provide limits for radioactive releases for radioisotopes in various forms. Some disposal of radioactive waste can be conducted through sanitary sewerage systems and licensed public waste disposal systems, if certain criteria are met. The US EPA has developed rules for air and water emissions of radioactive material. The controls on water releases have been in effect for a long time. The EPA and NRC have agreed on exposure limits caused by radioactive air emissions—promulgated in the National Emission Standards for Air Pollutants (NESHAPS) standard. Many states are in the process of adopting/adapting the NESHAPS standards or have already adopted them. As evidence of compliance, EPA relies heavily on computer software to model the emissions from a given facility and show that no person (workers or public) will be exposed above the limit of exposure. The limits vary according to the type of licensee or facility. The modeling software is available at no charge.

Environmental monitoring methods are usually of two types: (1) obtaining grab samples and (2) performing continuous monitoring. *Grab sampling* is exactly as it sounds. One simply uses a container to "grab" a portion of the media to be sampled and then analyzes the sample in the field or returns it to the lab for processing. *Continuous monitoring* consists of setting up a device or devices to measure radioactivity at a given point over time. It may be a system of measuring radiation directly, or one of obtaining periodic samples and determining radioactive content. Both passive and automatic systems may be employed. Grab samples only provide information for the point in time that they were collected. Continuous monitoring is usually more informative (and preferred more), but the systems are usually expensive and require frequent maintenance.

Airborne Emissions

Particulate, Gaseous

Emissions and contaminants can be caused by normal operations or by accident. To monitor these, some of the more common environmental monitoring methods consist of the following methods.

Air emissions may be releases through exhaust systems (which should be filtered), "open work" with unsealed radioactive materials or caused by fire or volatile chemicals. The form may be particulate or gaseous.

To measure airborne *particulates*, a pump with a sample filter is set up and operated for a known period of time. As the air is pumped through the unit, the radioactive particles are deposited on the filter and the air flow is measured (total volume of air). The radioactive content of the filter is measured and, using the volume of air pumped through, the concentration of radioactive material in the air is determined. The concentration (Curies per unit volume of air) can then be used to compare against regulatory limits. Some systems allow nearly continuous monitoring through automated sample changers and automatic measuring and recording devices.

For *gaseous* samples, a sample of air can be obtained in a closed container and the container contents measured. This can be done for Radon measurements where the volume of air is that of the container and the container is also part of the detector—having a lining of scintillation material which allows counting of the photons emitted. Another method is to process the air through a chemical solution or a solid material (such as activated charcoal) to capture the radioactive contents for analysis.

There are several systems which can directly measure the radioactivity content of the air in which they are placed. Air sampling generally can be used to measure alpha, beta, gamma, and neutron emitters—but the methods usually must allow for naturally occurring Radon daughters.

Virtually all systems of determining the *soil* content of radioactive *contaminants* involve obtaining grab samples. Samples can be taken from the surface or from measured points below grade, although there are some circumstances of radioisotopic soil contamination where the concentrations can be "estimated" by measuring radiation levels above the surface at a given distance. Once a grab sample (typically 500 to 1500 grams) has been obtained, the radioactive content can be determined by analyzing the radiation emitted with a scintillation or detector or by chemically extracting the material from the soil and then analyzing in a similar manner. The techniques and procedures used in soil sampling may also be applied to contaminated solids (such as industrial debris). Common interferences are natural radioactivity from Radium and its daughters as well as Potassium-40. Alpha emitters are probably the most difficult to evaluate since a chemical method is usually required. The regulations specify limits of contamination, but alternate (higher) limits can be applied for (application to the agency) if it can be shown that the contaminated soil will not cause a person to be exposed above any limit. This may occur when a facility is seeking to release an area for "unrestricted use" and finds that it cannot economically be decontaminated to below the stated limits. Usually, such an application should consider release by water runoff and by suspension (or re-suspension) of soil contaminants in the air. Soils often become contaminated by exposure to contaminated fluids or by settlement of contaminated particles in the air.

Water Contaminants

It once was more common practice to release radioactive fluid wastes into bodies of water. For short lived radioisotopes, this was not usually a very great problem or hazard. However, the criteria for releases have become more stringent in the past decade (public sentiment being more of a driving force than actual radiation risks). Monitoring is usually conducted routinely to assure compliance during authorized releases and to provide for detection of accidental release.

Water sample processing and analysis is very similar to soil sampling—simply a different media. Grab samples are obtained as known volumes and measured directly or analyzed following chemical or filtering extraction to determine the concentration. There are automated systems which provide continuous monitoring by obtaining and analyzing small periodic grab samples as, or from, a continuous stream. There are also analytical systems which allow for automatic, rather easy, analysis of liquids containing radioisotopes—such as liquid scintillation. Most types of radioisotopes can be readily analyzed—with naturally occurring Potassium-40 and Carbon-14 being the primary interferences.

It should be noted that the EPA has issued standards for limiting radioactive content of drinking water and the methods used in analysis of drinking water are generally the same as those touched upon here.

Surface Contaminants

As previously indicated, surface contaminants occur as loose or fixed radioisotopes. Fixed contaminants usually can only be evaluated by using a survey instrument with a probe/detector appropriate for the type of radiation involved. The instrument should be calibrated with a known source under the same geometry (distance from surface to probe). Usually, the radioisotope must be known to perform a proper analysis. Loose contaminants can be wiped with a suitable material (filter paper, wet or dry cloth wipe, Q-tip, etc.) and then the wipe can be taken to a lab or remote area for analysis. The type of analytical equipment used will be determined by the radioisotope and its emission(s).

Surface contamination limits are specified by concentration per unit area. Typical limits are number of dpm per 100 cm^2, 200 cm^2, or 500 cm^2. The limit may specify an average over the entire surface area of concern (such as a shipping container) with no single area to exceed some higher limit. This may require multiple sampling of a given object. A rule of thumb is to make a "Z" shaped wipe, without lifting it, with about 4 inches on each arm of the "Z". This covers most of an area roughly 10 cm by 10 cm (100 square cm). Sticky tape pieces with dimensions of the desired area can also be used. Of course, the person performing the sampling should be wearing protective gloves and should be prepared to place the wipe in an envelope or plastic bag.

Radiation Levels

Generally, measuring "free air" radiation levels is the easiest of the various monitoring tasks. Using calibrated survey instruments (and recording the results) is the simplest method. Other methods include placement of continuous monitoring instruments, film badges, or TLDs at critical or representative points.

> *Note: Sample identification and conditions of collection should be documented at the time of sample collection (or at the beginning and end of collection). Further, documentation of the analysis of the sample and the results should also be documented. These documents will provide evidence of compliance with the requirements. While they may also document non-compliance, identification and correction of radiation problems will be far more acceptable than ignoring or overlooking a problem.*

When establishing a monitoring program, careful consideration of the potential radioactive contaminants, the site characteristics, and the monitoring techniques and systems available will help determine the ease of implementation and an acceptable cost. One should keep in mind that there are many parameters that may Across over and cause interference—or they may provide for a simpler monitoring system. Soil sample collection points can be selected to indicate water and air contamination problems. Water samples can provide for measurement of soil and air contaminants. Air samples can indicate soil contamination. Radiation levels can help to evaluate airborne and soil contaminants. Few systems are truly isolated.

Summary

Although the full risks associated with using radiation are not always known, the worker's concern about his/her health effects can be mitigated by providing proper radiation protection tools. Further, the health and safety of man and the environment can be protected from radiation hazards by maintaining proper controls over radioactive material and by properly monitoring radiation hazards to assure effective controls are maintained.

Chapter 4

Safe Operations Using Sources of Radiation

1. Introduction

This chapter introduces methods and procedures for using, handling, storing, and transporting radioactive materials. Some of the procedures are also applicable to use of other types of use of radiation sources.

2. General Radiation Operations

Control of Radiation Sources

Radiation control regulations are designed to require authorized users of radiation sources to control their sources to the extent necessary that will prevent unnecessary or harmful radiation exposures of workers and members of the general public. This control extends to the release of radioactive materials into the environment. Loss of control may lead to administrative or civil penalties, particularly if negligence or willfulness is determined. *Common sense* should be a major ingredient in designing control systems and procedures.

A person must possess a valid radioactive material license issued by a radiation control agency BEFORE acquiring and using radioactive material (there are a few exceptions). When applying for a radioactive material license, the agency will require submission of procedures that assure proper controls over radioactive materials will be in effect. Most such controls are specifically required by the regulations for industrial radiography and well-logging applicants, but the controls for other users are arranged through written procedures which are, in effect, approved by the agency by issuance of the radioactive material license. Once approved, they have the same effect as regulations and law.

For x-ray equipment (machine produced radiation), registration of the use is required—not a license. In most cases, an application for registration must be submitted to the agency within 30 days following the first use (not installation). Persons or companies using devices that require special facilities and shielding, such as accelerators, should submit their plans before beginning construction to avoid the expense of "retro-fitting".

Methods of Control

Security Measures, Fire Controls, Physical Controls, Records, Training

Maintaining control of radiation sources at all times will help to prevent accidents and injuries. When any source is being used, stored, transported, or having maintenance performed, control must be constantly maintained. Reviewing "standard" control methods and carefully applying them to specific operations, while providing *appropriate* modifications, will provide a good level of security and safety. Often, a licensee (person/company holding a radioactive material license authorizing use/possession of radiation sources) will simply adopt others' procedures without considering the effects on his/her particular program. This usually results in unnecessarily stringent self-imposed requirements causing non-compliance in areas not really applicable, or it results in insufficient safety and control procedures with the potential for causing accident or injury. A safety program and its controls should be specific to the particular aspects of the use of radiation sources by the licensee.

Security Measures

Proper *security measures* are essential in maintaining control of radiation sources. The regulations require that radiation sources not be removed by persons not authorized or qualified to deal with them. It is essential to determine personnel control measures and appropriate procedures for locking of areas, buildings, rooms, vehicles, and source containers that may have radiation sources located within them. The security controls need not be absolute—simply appropriate. If one uses a good padlock on the door of an enclosed room of standard construction to store a small source, say 10 mCi of Cs-137, it is probably appropriate. If the source happens to be a 100 Curie Ir-192 industrial radiography source, then such storage may not be appropriate. Since the latter source can cause far more immediate physical damage to a person handling it when unshielded, more stringent security is reasonable. The author recommends at least three "[18]levels of security" should be provided for smaller sources and at least four

18 *Note: The author prefers to use the concept "levels of security"—each device/method designed to prevent access being one level. For example, a source stored in a locked container, in a locked room, in a locked building in a locked fenced area, with a security guard on duty 24 hours per*

levels for sources that can immediate, serious harm when handled in an uncontained manner—such as in the hand or in direct contact with a person's skin.

Fire Controls

In addition to security controls, each program should incorporate fire controls. Many cities require that hazardous materials (which includes radioactive materials) be registered or reported to the local fire department. The fire department will usually send a representative to obtain details on the hazards so that fires and emergencies can be handled safely by fire fighting personnel. Should a unit respond to a fire or emergency and then learn that radioactive materials are stored within, not knowing the location and degree of hazard might cause the unit to provide minimal fire-fighting or emergency assistance. On the other hand, a fireman might enter an area that could be truly hazardous due to the presence of radioactive material. Proper planning and information exchange will minimize the chance of improper handling of an emergency by fire and emergency response personnel.

Physical Controls

Occasionally, an agency will receive a report of a missing radiation source. It may have been lost or stolen—or it may be missing because of vandalism (often caused by a former employee that left the company under strained circumstances). On the other hand, it may be present or in use, but the licensee happened to have many sources and was unable to identify the particular source or its location. These circumstances can be prevented or mitigated by having proper *physical controls* incorporated into the program. Physical controls include measures such as periodic physical inventories; procedures for removing sources from storage, using them, and then returning them to storage; and performing equipment inspections at the time of use, transport, and storage. Periodic inspection and maintenance of devices containing radioactive material also fall into this category.

Records

Records documenting radiation source information and conductance of safety procedures have several important functions in the safety process. As a minimum, records enable the regulatory agency to verify that all requirements are being met (an agency is, after all, responsible to the public), allow inspectors the opportunity to identify and point out potential safety hazards, allow the licensee/registrant to audit and evaluate their safety program, and have been instrumental in evaluating accidents. Poorly designed or maintained records and incomplete records have

---

day, would have 5 levels of security. Personnel in attendance would be one level, as would an operable intruder alert system.

contributed to many compliance problems. If a record (most records are pre-designed forms) is too complex and/or difficult for workers to consistently provide the proper information, then the record will probably be unsuccessful. Duplication of information should be avoided. As many as four or five records have been observed being used in one program to document, essentially, one function. This increases the likelihood of occurrence of incomplete records and wastes a considerable amount of time for both the program and the agency inspector (during inspections). (See the notation about government records in Chapter 5.)

Training

Training of personnel is, perhaps, the best single approach to preventing accidents and injuries in working with radiation sources—or any hazard. In most uses of radiation, training for radiation safety officers and radiation workers is required. The training program content may require training programs of 4 to 40 hours, or more. A licensee/registrant may implement his/her own training program, having it approved during the application process. Since setting up and providing proper training requires time and resources, many operations choose to send their staff to agency approved or agency accepted training courses (which may be given "in-house"). An agency approved training course is one that has had the training procedures and materials reviewed by agency staff and an agency staff member has evaluated the course by attending it. An agency accepted course is one that has had the training procedures and materials submitted, reviewed and accepted by the agency—but there is no in-class evaluation of the course. Currently, industrial radiography and well-logging licensees are required to provide approved training courses. Further, certain medical uses have more stringent training and experience requirements in both the use of radioactive material and the use of x-ray devices.

Control Procedures

Physical Inventories, Use and Storage Procedures, Inspection and Maintenance

There are a number of *control procedures*, common to many radiation use programs, that need to be considered when implementing a good radiation safety program. In some cases they are required. In others, they are recommended or unnecessary.

Physical inventories

Physical inventories should be performed every six months, as a minimum, if a licensee possesses a number of sources with sporadic use or if sources are transported to other locations on site or to other sites. Such inventories are generally required through commitments made by the licensee when applying for, or upgrading, a license, or when adding new authorizations to a license. Texas regulations require that industrial radiography and well-logging licensees perform physical inventories every three (3) months—or quarterly. This should be the standard throughout the United States.

Use and storage procedures

Use and storage procedures must consider the entire process. Typically, a radiation source is removed from a storage facility, taken to a work area for use, and then returned to the storage facility. Control measures should not only account for EACH STEP of this process, but also for any VARIATION that might occur. Most control procedures assume that an authorized employee will be present during all steps, but overlooking a procedure of temporarily securing a source when the employee must leave the area momentarily can lead to accident, injury, or loss.

Equipment inspections

Equipment inspections at time of use, transport, and storage and more extensive *periodic inspection and maintenance* are essential in preventing radiation hazards and problems. Many accidents have been found to be caused by faulty equipment—either safety or non-safety devices. For example, industrial radiography sources are kept in a special, shielding container called a "camera". To use the source to obtain a radiograph (film exposed by radiation that has passed through an object placing an image on it), the source must be pushed from the shielded position within the camera to the object to be radiographed. A cable is attached to a connector and a cranking device (called a "crank-out") is used to push the cable through the camera, which in turn pushes the attached source out of the camera through a guide tube to the point where the radiograph will be performed. This system allows the radiographer to maintain a maximum distance from the source while it is out of the camera. After the radiograph has been performed, the crank is cranked in the opposite direction to pull the source back into the camera. The cranking mechanism is not necessarily a radiation safety device. In itself, it doesn't provide radiation shielding or measure radiation levels. However, if it does not work properly, the source may not be returned to the safe position, thereby creating a radiation hazard. At one time, the condition of the crank-outs was not considered in the regulatory process and an agency's authority to require corrective action was questionable. An inspector would often see one or two of the four or five screws, which hold the crank-out body together, missing. A number of accidents investigated indicated that missing screws were a cause of jamming of the crank-outs. While the crank-outs may not be safety devices, they are obviously "safety related". The regulations were modified to address requirements for inspection of equipment related to radiation safety for certain types of use and the number of accidents dropped. For types of use not having safety related devices specifically addressed in radiation control regulations, inspection and maintenance procedures for safety related equipment is usually addressed in license (or registration) conditions and/or in written safety procedures.

Storage Precautions

When radiation sources are not in use they must be safely stored to prevent any accident which might cause unnecessary, even harmful, exposures to radiation or radioactive material.

Safely stored radiation sources will prohibit the unauthorized removal by persons not properly trained and equipped to handle them. Safe storage includes implementation of both proper shielding and security measures.

Prevention of Exposure and Unauthorized Removal

Shielding & Distance, Physical Security

As we learned in Chapter 2, *shielding* materials can be used to reduce radiation exposure. Providing for distance can also be used to keep radiation exposures in check when shielding may be a costly proposition. Most storage facilities provide both shielding and distance to assure that radiation levels in unrestricted areas meet regulatory limits. *Distance* is used by providing a physical barrier, such as a fence or wall, which separates the unrestricted area from the storage area. Distances can easily be calculated or determined by measurement. However, one should consult with a person qualified and/or experienced in shielding materials and methods in order to optimize shielding at a minimum cost.

A storage facility must also provide *physical security* to prohibit unauthorized persons from tampering with stored radiation sources. In addition to assured radiation safety, proper security will help avoid the additional expense to a company of replacing expensive equipment—or decontaminating an area caused by tampering (decontamination can run into the millions of dollars). An x-ray storage facility need only provide security since radiation is produced only while the unit is operating. However, if the storage facility doubles as a use area, then shielding may be required. Security for storage of radioactive materials will be discussed more thoroughly in Chapter 7.

There are some terms and devices commonly found in the "tech talk" of radiation safety. A "storage bunker" or "storage safe" are terms often used to describe the primary storage component. A typical storage bunker may be constructed of strong containment material, such as steel, for security—with shielding added within the bunker or around its perimeter. The bunker may fixed down into the ground or concrete slab, or may be set on top. It may be separated from, or fixed to, a building. In some cases, particularly well-logging, a bunker may be constructed by seating one or more large diameter pipes down into the ground or slab. This is called "down-hole" storage and allows the soil below grade to provide shielding when the sources are lowered into them. This is a cost effective method used in storage of large gamma and neutron sources used in the well-logging industry. When designing a storage facility, one should consider the procedures involved in storing and removing sources. If sources must be removed from their shields to place them in the storage facility, such as in down-hole storage, then personnel will receive additional exposure during the process.

Many radiation sources are kept in operational devices that also provide adequate shielding. The devices for some sources do not provide adequate shielding for storage, so additional shielding must be considered for the storage facility. For example, moisture density gauges have "smaller" sources in well-shielded cabinets which are integral components of the instruments, and storage in their transport containers in a locked closet or room should meet requirements. However, a company possessing a large number of the gauges may need to consider shielding or distance to assure radiation levels are not excessive. On the other hand, radiography cameras provide adequate shielding for all functions except storage. A well-constructed storage bunker with additional shielding will probably be required—particularly if a number of cameras will be stored. A down-hole bunker may be a good option for the storage of radiography cameras. Storage facilities should be designed to allow for the addition of more sources should the company expand its operations. It should be noted that storage facilities, once approved by an agency, cannot be changed without approval by the agency prior to making the change(s).

Handling and Disposal of Radioactive Waste

Regulations[19] define radioactive waste as "*Any discarded or unwanted radioactive material, unless exempted by agency rule or any radioactive material that would require processing before it could be put to a beneficial reuse*". There are a few exceptions. The licensee that produces radioactive waste is referred to as the generator.

Radioactive waste (often "rad waste") has been a very sensitive issue with the general public. Many persons appear to believe there is a much greater threat from a vehicle in transit on a public highway and carrying a number of 55 gallon drums containing

19 TAC §289.201(b)(82)

laboratory generated radioactive waste than from an industrial radiographer's vehicle in transit on the highway. Since they are unknowledgeable in the physics and procedures of radiation matters, they don't realize that the radioactive waste perhaps totals a half of a curie while the radiography vehicle is carrying 100+ curies of Ir-192 (which can cause serious burns from a brief touch of the source). Therefore, it is both radiologically and "politically" expedient to establish sound procedures in dealing with radioactive waste. There are two major areas to consider: handling radioactive waste as it is generated and stored and proper disposal methods when disposal becomes necessary. Although there are a number of classifications of radioactive waste, for simplification here we will divide them into two principle types of waste: unleaking sealed sources and loose materials.

Radioactive waste as sealed sources that are not leaking can be handled with the same precautions as a useable source and stored in the same manner (proper shielding and security). However, dealing with unsealed radioactive wastes can be extremely difficult at times. For normal operations, procedures should be established to identify, classify, containerize, and safely store generated wastes. If the wastes are volatile (experience evaporation into the air) they may need to be stored in units that control air flow—such as fume hoods. Handling waste containers requires use of the same protective procedures as used for handling "stock or raw material and product" or contaminated materials (protective clothing, gloves, shoe covers, etc.). If storage of stock material and performance of work must be confined to a fume hood, storage of properly labeled waste containers in a fume hood is probably also necessary or desirable. However, bio-hazardous and chemo-hazardous properties of waste, caused during processing or use, should be considered. It may be necessary to provide completely separate, yet equal, storage facilities.

Disposal Methods and Resources

The manner of classifying and containerizing radioactive waste (segregation) should consider the anticipated method or mode of disposal. Short lived materials in small quantities may be disposed through the sanitary sewerage system under certain circumstances, and others that qualify may be disposed at licensed disposal facilities. Some may only be disposed at a licensed waste disposal facility (a limited and costly disposal resource). Generally, solid wastes are easier to work with than liquid wastes, so one may need to consider a step of solidifying liquids before containerization. Carefully established waste handling procedures and proper training of workers in those procedures will save any operation time, money, and possibly embarrassment. Further, it is important to assure that rad waste is not placed into non-radioactive (normal) waste containers as, in addition to the radiation hazard presented, agency inspectors do survey normal waste streams. Nor should normal waste be placed into rad waste containers—since the cost of rad waste disposal is usually much higher and volume is a major factor in the cost.

There are number of disposal resources. They consist of:

- Return to supplier
- Licensed disposal facilities
- Waste processors and brokers
- Sewerage systems
- Licensed municipal disposal facilities
- Treatment by incineration
- Normal waste streams

During the license application process, the applicant should identify/designate a waste disposal resource. At the time disposal is actually performed, other resources may be used. Some options are briefly addressed below.

Return Waste to Supplier

Disposal by returning waste to the supplier is the simplest approach to radioactive waste disposal. One should arrange with the supplier of the source or material to take it back. A written commitment (preferably a contract) should be the method of agreement, if possible. The cost of paying for return shipment to the supplier by the licensee will probably be much less than the cost of disposal at a licensed disposal facility or using a broker. Most likely, suppliers will have already folded associated disposal costs into the prices of their products and adding a little more waste may not add much to their anticipated expenses.

Ship Waste to Licensed Disposal Facility

Disposal by shipping waste to a licensed disposal facility is an option that is mostly viable only for large companies with many resources. In addition to the high charges of disposal, disposal facilities often require indemnification on the order of a million dollars. Few companies can afford the shipping charges, disposal fees, and indemnification. Failure to properly fill, seal, label, and transport containers can lead to penalties by the state (regulating the disposal site) and restriction and/or additional charges by the disposal facility.

Dispose of Waste through Licensed Waste Processor or Broker

Disposal through use of the services of a licensed waste processor or broker is an option to consider. A number of service companies possess radioactive material licenses that allow them to collect customers' wastes, process them to properly segregate and containerize them, and then transport them to a licensed disposal facility to be disposed—while providing the necessary indemnification. This option is used by most companies with modest resources. However, it should be noted that the original generator may possibly retain responsibility for the wastes should future legal problems arise.

Dispose of Waste in Sanitary Sewerage System

Disposal in sanitary sewerage systems can be considered as the regulations (state and federal) have provisions for disposal of certain radioactive wastes by flushing them into sewerage systems—WITH limitations. The Texas regulations[20] define sanitary sewerage as: "A system of public sewers for carrying off waste water and refuse, but excluding sewage treatment facilities, septic tanks, and leach fields owned or operated by the licensee or registrant". Discharge of (waste) radioactive material by release into sanitary sewerage is authorized under the following conditions:

(1) the material is readily soluble, or is readily dispersible biological material, in water;

(2) the quantity of licensed radioactive material released into the sewer in 1 month divided by the average monthly volume of water released into the sewer by the licensee does not exceed the concentration listed in the table given in the regulations;

(3) if more than one radionuclide is released, the sum of the fractions (as monthly averages) for each must not exceed unity, and

(4) the total quantity of licensed radioactive material released into the sanitary sewerage in a year does not exceed 5 curies (Ci) (185 Gigabecquerels (GBq)) of hydrogen-3, 1 Ci (37 GBq) of carbon-14, and 1 Ci (37 GBq) of all other radioactive materials combined.

It should be noted that radiation control regulations allow that "*Excreta from individuals undergoing medical diagnosis or therapy with radioactive material are not subject to the (above) limitations . . .*". In other words, this category of waste can be placed into sewerage systems without controls.

---

[20] This discussion is based on the Texas radiation control rules [TAC §289.202(gg) February 2009] found at http://www.dshs.state.tx.us/radiation/rules.shtm. The state or federal rules applicable to your program should be consulted and followed.

Discharge by release into septic tanks is prohibited by radiation control regulations[21], which state: *"No licensee shall discharge radioactive material into a septic tank system except as specifically approved by the agency"* (this, of course, is highly unlikely). Sewerage systems allow for dilution to reduce the concentrations—although the remaining materials will ultimately be reconcentrated at the end sewerage treatment plant. Septic tanks simply reconcentrate the material and potentially create a radiation hazard.

To use the sewerage system disposal option for disposal of a single isotope, first ensure that the material is soluble, or for biological material, will be dispersible in water. Then determine the average water released into the sewerage system by using the water bill or obtain the information for the building lease company. Calculate the average concentration by dividing the monthly concentration by the monthly average water amount (a conversion from gallons to liters will probably be necessary). Compare the estimated concentration with the table value for that isotope. Ensure that the total release will not exceed 5 curies in a year. If the concentration exceeds the limit by a large amount, it will be necessary to consider alternatives. Decay may be possible for short-lived isotopes. Perhaps all or a portion of the waste could be collected by a broker service. Waste and disposal records should document these activities to demonstrate to the agency that the requirements were met.

For disposal of multiple isotopes, the steps are the same, but the contribution by each isotope must be considered (which calls for a more complex calculation of concentrations) and the total of all isotopes cannot exceed 5 curies per year. For example, for disposal of two isotopes, say isotope A and isotope B, we would first calculate for each the monthly concentration that will be released. We then calculate the "fraction" of the limit for each isotope being released by dividing the release concentration by the concentration limit. Suppose isotope A will be 1/10th (or 10%) of its release limit and isotope B will be 1/5 (or 20%) of its (different) release limit. Together, they cannot exceed unity (sum to over 1). Summing the two fractions yields 1/10 + 1/5 = 1/10 + 2/10 = 3/10. Since 3/10ths is less than 1, the release would meet the requirements. (Of course, actual calculations would use exponential form for convenience—the calculations need not be shown as "fractions".)

Disposal at Licensed Municipal Facilities

In Texas, and other states, disposal at licensed municipal disposal facilities is a possible option. Through exemption, the Texas regulations[22] allow the disposal of small quantities of some isotopes meeting certain, specific criteria when using

21 TAC §289.202(ii) [March 2006]

22 TAC §289.202(fff) [February 2009]

specific disposal procedures. The number of these isotopes in Texas has been greater than those authorized by other states or the NRC. This category has been commonly termed below regulatory concern, or **BRC**, waste. Currently, prior agency approval for specific disposal is not required, but the procedure is usually addressed in the license application process. The disposable isotopes and their forms (or media) common to all states and the NRC are (quoted):

- 0.05 microcurie (µCi) (1.85 kilobecquerels (kBq)), or less, of hydrogen-3, carbon-14, or iodine-125 per gram of medium used for liquid scintillation counting or *in vitro* clinical or *in vitro* laboratory testing; and
- 0.05 FCi (1.85 kBq), or less, of hydrogen-3, carbon-14, or iodine-125, per gram of animal tissue, averaged over the weight of the entire animal.

The (Texas) regulations state: *"A licensee may discard the . . . licensed material without regard to its radioactivity"*. They do not say that the material is not radioactive or "deregulated" at this point. It remains radioactive waste and, while its radioactivity can be disregarded for purposes of disposal, it must be handled as radioactive material until the disposal has been completed. One must keep in mind that improper disposal can lead to high monetary penalties. The exemption only applies when all criteria have been met and the actual disposal has been accomplished.

The usual method of disposal is to take the waste to a licensed municipal landfill, although incineration (which requires advance agency approval) is occasionally used—particularly when there is a biohazard associated with the disposal. The waste may also be injected into hazardous waste well streams. Additional criteria are:

- Tissue may not be discarded in a manner that would permit its use either as food for humans or as animal feed; and
- Records of the disposal(s) must be maintained.

Records of such disposals should include, as a minimum, a description of the wastes (isotopes, concentrations, form), dates and times of disposals, locations of disposals, and identification of the person completing the records.

Prior agency approval is required for the additional isotopes allowed to be disposed in Texas. In the early 90's, Texas licensees and interested parties petitioned the TDH for rule making to allow more liberal disposal of radioactive waste which represented a low radiation hazard. The petition was successful and the rules were modified to permit disposal of (popularly termed) "300 day half-life" isotopes that meet specific criteria. Upon agency approval, this rule[23] allows:

- Any licensee may discard (listed) licensed material provided that it does not exceed the (listed) concentration and total curie limits contained therein;
- Disposal in a municipal solid waste site[24] unless such licensed material also contains hazardous waste[25].

Conditions for disposals are:

- surveys, adequate to assure that the limits are not exceeded, must be performed and recorded; and
- all labels, tags, or other markings that would indicate that the material or its contents is radioactive must be removed or otherwise obliterated or made obscure.

Application for an authorization requires the submission of procedures which describe:

- the physical delivery (mode, methods) of the material to the disposal site;
- surveys to be performed to show compliance with such requirements;
- methods of maintaining secure packaging during transportation to the site; and
- methods of maintaining records of all disposals.

Approval of this method does not exempt the authorized party from other rules and requirements. The Texas rule[26] specifically states: "*Nothing (in this section) relieves the licensee of maintaining records showing the receipt, transfer, and discard of such radioactive material*". Further, the Texas rules[27] also require: "*Nothing (in this section) relieves the licensee from complying with other applicable federal, state, and local regulations governing any other toxic or hazardous property of these materials*". Therefore, all radioactive materials being "discarded" must be maintained controlled throughout the process—until finally disposed—and none of the materials can be mixed with or contain other materials that are prohibited from disposal in such a landfill (such as

23 TAC §289.202(fff)

24 Type I municipal solid waste site, as defined in the (Texas) Municipal Solid Waste Regulations of the authorized regulatory agency (31 TAC Chapter 330).

25 Hazardous waste, as defined in Section 3(15) of the Solid Waste Disposal Act, Health and Safety Code, Chapter 361. Any licensed material which is a hazardous waste as defined in the Solid Waste Disposal Act may be discarded at a facility authorized to manage hazardous waste by the authorized regulatory agency.

26 TAC §289.202(fff)(7) [February 2009]

27 TAC §289.202(fff)(8) [February 2009]

bio-hazardous or chemo-hazardous waste that are specifically prohibited! *Caution! In Texas, agency staff have surveyed waste ready for such disposal and have performed surveys at disposal sites—leading to at least one case of civil penalties for violations!*

Disposal Through Treatment by Incineration

Disposal by incineration is generally not available for radioactive waste disposal. However, treatment by incineration is not necessarily prohibited. The Texas rules[28] state *"A licensee may treat licensed material by incineration only in the form and concentration specified in . . . [applicable rule] . . . or as authorized by the agency"*. The materials (described in the previous section) can be disposed by incineration—if it is incorporated into the application for authorization and is approved by the agency. This may be a viable option for contaminated human and animal tissues.

Disposal in "Normal" Waste Streams

Believe it or not disposal in "normal" waste streams is possible. However, each licensee must insure that his/her procedures are sufficient to prevent intentional or accidental placement of radioactive materials in the normal wastes—with one notable exception. Exempt sources, such as smoke detectors, can be disposed in normal waste. However, DO NOT assume radioactive material is "exempt" because of its small quantity! Such material must meet specific criteria to qualify as exempt.

Summary of Disposal Methods

Radioactive waste disposal resources have tended to decrease and then increase in a cycle dictated by varying political and public sentiments. One factor has remained constant, though. The disposal costs continue to rise higher and higher. The authorizations allowing disposal of short lived radioactive wastes through local resources (such as the municipal landfills) has not only kept disposal costs lower, but have reduced the volume of wastes containing radioactivity of little significance being sent to licensed disposal facilities. By reducing this volume, more capacity is made available for disposal of wastes that DO need to be disposed in a properly engineered waste disposal facility.

Radiation Surveys

Since there have been many references to radiation surveys, a review of some of those commonly required or used in various operations using radiation sources is in order.

---

28 TAC §289.202(hh) [March 2006]

Surveys for External Radiation Levels

Storage Area, Use Area, Vehicle, Device/Container, Package Receipt, Records

It is clear that the performance of radiation surveys at key points during operations maintains a high level of safety and helps to avoid embarrassing compliance problems. Documents of periodically performed surveys have enabled radiation safety officers and agency inspectors to identify and minimize potential hazards. Although few uses/programs will be required to perform most of the possible types of surveys, the ones discussed here will be used most frequently by many programs.

Storage Area Surveys

Storage facilities are designed for long-term storage of radiation sources while they are not being used. Once set up, it is assumed that stored sources are safely "out of the way". *Storage area radiation surveys* are performed to confirm that radiation levels around the facilities meet regulatory limits. Once a survey is performed, it should be valid until the storage conditions change. If periodic surveys are required, the survey would be valid until the next survey is due. Changes in numbers of sources stored, container shielding, facility shielding, controlled areas around the facility, and source positioning within the facility can all invalidate a recorded survey as each of these may cause changes in the radiation levels around it. In general, a storage area/facility survey should be performed periodically (each week, month, six months, or year) and then resurveyed when one or more of the storage conditions changes. It should be noted that the frequency of required periodic surveys will be set by specific requirements of the regulations or by the licensee/registrant's agency approved operating procedures. A storage facility with few sources that are used infrequently may need only be surveyed every six months or year. A facility with a large number of sources with very frequent check-out and check-in of sources may need to be surveyed at the end of each day or work period. Records should indicate the radiation levels, usually in mr/hr (neutron flux should be converted to units of rem), at the confines of the actual storage facility, at each near-by work station, and at the boundaries of the restricted areas. The highest measured radiation level should be recorded for each wall or surface. The height of the measurement should represent the "center point" of the average person that might be standing in the area, i.e. about waist to chest high.

Use Area Surveys

Use area radiation surveys are performed to show that radiation levels at the boundaries of established control points/barriers for restricting the area(s) to protect against radiation exposure will not exceed limits (discussed in Chapter 2) for exposures in unrestricted areas. They will also show the potential exposure of other employees or workers, not involved in radiation procedures, that may "happen" into the area.

The survey records can be used to estimate inadvertent exposures. In some field operations, such a survey may not be required because of practical limitations and in view of the fact that special circumstances reduce the likelihood of inadvertent exposures. For example, in the well-logging industry where large neutron sources are used (1-20 curies), field site surveys are not necessary because the source is only un-shielded for a few minutes, neutron surveys are difficult, and gamma surveys are not very effective for neutrons. However, for tracer studies, where radioactive material is injected into oil/gas wells, post operation surveys are useful for determining whether contamination is present on the site.

Vehicle Surveys

Vehicle radiation surveys are generally only necessary when a vehicle may be used for overnight storage when conducting operations away from the main licensed storage facility. When DOT transport requirements are met and transport conditions are always the same (container radiation levels, storage position on vehicle, etc.), day-to-day surveys are not very necessary. When the only safe overnight storage mode is by storage on the vehicle in unrestricted areas, then it is necessary to show that radiation levels around the vehicle will not cause any excessive exposures to members of the public. Some uses do require vehicle surveys for routine transport. The license applicant should determine specific requirements and needs during the application process and incorporate them into the "operating and safety procedures".

Device Surveys

Device/container radiation surveys are performed in some operations to verify that sources are safely stored in the device or container, shutters are closed, shielding is intact, external radiation levels meet requirements (where applicable), and labeling of the container and posting of the area(s) of use or storage is appropriate. For example, radiation control rules[29] state: "*A room or area is not required to be posted with a caution sign because of the presence of a sealed source(s) provided the radiation level at 30 centimeters from the surface of the sealed source container(s) or housing(s) does not exceed 0.005 rem (0.05 mSv) per hour*". Obviously the container/device will need to be surveyed and the results recorded to verify that the radiation level is less than 5 millirem at 30 centimeters (11.8 in) if the room or area will not be posted with a caution sign.

---

[29] TAC §289.202(bb)(3) [October 2000]

Package Receipt Surveys

Package receipt radiation surveys are required to be performed and recorded when packages are received. Texas rules[30] require each licensee, upon receipt of a package containing radioactive material, to monitor:

- the external surfaces of a labeled package, labeled with a Radioactive White I, Yellow II, or Yellow III label . . . for radioactive contamination unless the package contains only radioactive material in the form of gas or in special form . . .; and
- the external surfaces of a labeled package, labeled with a Radioactive White I, Yellow II, or Yellow III label . . . for radiation levels unless the package contains quantities of radioactive material that are less than or equal to the Type A quantity . . .; and
- all packages known to contain radioactive material for radioactive contamination and radiation levels if there is evidence of degradation of package integrity, such as packages that are crushed, wet, or damaged.

Records of the surveys and monitoring are currently required to be maintained for agency inspection for three (3) years. Leakage of radioactive contents or excessive radiation levels are required to be reported to the agency.

To monitor packages for leakage, one might use and analyze wipes taken from the package surfaces. A simple approach is to examine the package and, if there is no apparent leakage, remove the inner container (or, alternatively, the outer container or package) to another area and then survey the outer package. If leakage had occurred during transport, then the package would probably have an elevated radiation level. When removing the inner container(s) protective equipment (gloves, lab coats, etc.) should be worn and anti-contamination surface covers placed under and around the potentially contaminated inner and outer packaging materials and containers.

Radiation survey records generally must include:

- the date and the identification of individual(s) making the record;
- a unique identification of survey instrument(s) used;
- an exact description of the location of the survey;
- survey results; and
- appropriate units of measure.

It should be noted that records of receipt, transfer, and disposal of sources of radiation must uniquely identify the source of radiation. Most records requirements are described in a records section of the applicable rules.

---

30 TAC §289.202(ee)(2)(A) [October 2011]

3. Transport Requirements and Procedures

Transport Regulations and Authority

Transport requirements stem from three governmental authorities, the federal government's US Department of Transportation (USDOT) and US Nuclear Regulatory Commission and the local state government (if any rules or laws have been enacted). In Texas, for example, the Texas Department of of State Health Services (DSHS) has promulgated rules for transport of radioactive material. However, the Texas Department of Public Safety (DPS) and the Texas Department of Transportation (TXDOT) also enforce federal requirements. These same conditions will probably exist in most, if not all, of the Agreement States. In general, persons transporting radioactive materials in interstate commerce must abide by the federal provisions and persons in intrastate commerce by the state provisions. The latter would be, in actuality, licensees of a state radiation control program. In most cases, a state's transport regulations are "in addition to and not a substitute for" other existing rules. State licensees transporting radioactive materials in their own state will fall under both (federal and state) jurisdictions. The two sets of requirements are so similar that no conflicting requirements should be encountered. The DPS has adopted the US DOT rules and will most likely enforce those. Radioactive material is classified as a hazardous material and falls under many of the general provisions regarding transport of hazardous materials. It should be noted that a state agency is exempt from USDOT rules by those same rules.

Shipping Requirements

Classification, Packages, Labeling, and Transport Index (TI), Shipping Papers

For shipments of radioactive materials, three parties are involved: the shipper, the carrier, and the consignee (receiver). In general, the shipper is responsible for correctly classifying the radioactive material, properly preparing it for shipment, and preparing necessary records. If the shipper also transports his/her own packages, then he/she will also be responsible for fulfilling transport requirements. Radioactive material licensees will often be the shipper, the carrier (transporter), and the consignee as they take their radiation sources off-site for work. X-ray machines do not fall under transport requirements since they do not produce radiation when not being operated.

> *Note: This section will be limited to transport of "common" shipments of radioactive material. The more "exotic" ones, such as: fissile materials, radioactive materials mixed with hazardous chemicals, low specific activity*

(LSA) materials, etc., will not be discussed, nor will there be discussion of related packaging, such as "industrial packages".

Classification

The "first step" in preparation for transport of radioactive material is to *classify* it and determine the appropriate packaging. (Note: In many cases, most of the work was done when the supplier originally shipped the source(s) to the licensee.)

The radioactive material must be classified as either *normal form* or *special form*. According to USDOT rules, CFR § 173.403[31], *special form Class 7 (radioactive) material means either an indispersible solid radioactive material or a sealed capsule containing radioactive material which satisfies the following conditions:*

- *It is either a single solid piece or is contained in a sealed capsule that can be opened only by destroying the capsule;*
- *The piece or capsule has at least one dimension not less than 5 millimeters (mm) (0.2 inch (in)); and*
- *It satisfies the test requirements of §173.469.*

There are some exceptions listed in the rule for older encapsulated sources. Also, this rule defines normal form Class 7 (radioactive) material as "*Class 7 (radioactive) which has not been demonstrated to qualify as 'special form Class 7 (radioactive) material.'*"

Radioactive material is also classified as Type A or Type B by CFR § 173.403[32], as follows:

- *Type A quantity means a quantity of Class 7 (radioactive) material, the aggregate radioactivity which does not exceed A_1 for special form Class 7 (radioactive) material of A_2 for normal form Class 7 (radioactive) material, where A_1 and A_2 values are given in §173.435 or are determined in accordance with §173.433.*
- *Type B quantity means a quantity of material greater than a Type A quantity.*

Radioactive material is generally transported in a package. According to USDOT rules, CFR § 173.403[33], *package means the packaging together with its radioactive contents as presented for transport.* Packages are classified by "type". The types of packages listed in the TXDOT rules are excepted package, industrial package (Types 1, 2, and 3), Type

31 49 CFR 173 : Shippers—General Requirements for Shipments and Packagings [49 CFR 173] (Transportation)

32 Ibid.

33 Ibid.

A package, Type B package[also Type B(U) and B(M)], and fissile material package. The only packages discussed here will be Type A and Type B.

Also according to CFR §173.403[34],:
The USDOT definitions of packagings are:

- *"Packaging means, for Class 7 (radioactive) materials, the assembly of components necessary to ensure compliance with the packaging requirements of this subpart. It may consist of one or more receptacles, absorbent materials, spacing structures, thermal insulation, radiation shielding, service equipment for filling, emptying, venting and pressure relief, and devices for cooling or absorbing mechanical shocks. The conveyance, tie-down system, and auxiliary equipment may sometimes be designated as part of the packaging."*
    - **Type A package** is *"a packaging that, together with its radioactive contents limited to A1 or A2 as appropriate, meets the requirements of §§173.410 and 173.412 and is designed to retain the integrity of containment and shielding required by this part under normal conditions of transport as demonstrated by the tests set forth in §173.465 or §173.466, as appropriate. A Type A package does not require Competent Authority approval.*

    - **Type B package** is *"a packaging designed to transport greater than an A1 or A2 quantity of radioactive material that, together with its radioactive contents, is designed to retain the integrity of containment and shielding required by this part when subjected to the normal conditions of transport and hypothetical accident test conditions set forth in 10 CFR part 71.*

The designation A1 is the maximum activity of special form radioactive material permitted in a Type A package and A2 is the maximum activity of radioactive material not special form . . . permitted in a Type A package. These values are given in§173.435 or are determined in accordance with §173.433. Most packages are either Type A or B. Excepted packages are used for shipping small, innocuous quantities that meet certain criteria. They have few requirements, such as documentation and labeling of the packaging.

Type A containers are designed to prevent release of contents during normal transportation and in the event of minor accidents. They can be of any shape and can be designed as boxes or drums. Materials can be cardboard, wood, plastic, metal, etc. Documentation that the container has passed the required test must be maintained by the <u>shipper</u>.

---

[34] Ibid.

Type B containers are designed to prevent releases during serious or severe accidents. Type B packages must pass more rigorous required tests and documentation that the container has passed the required tests must be maintained by the shipper. A *Certificate of Compliance* (COC), issued by the NRC for most containers, should be maintained by the shipper. There are additional requirements for international shipments.

Following through the requirements to prepare a package for shipment and then transporting it is no easy task. As an example, several hypothetical sources to be shipped are chosen and examined against most of the basic requirements for packaging and transport in stages. The following isotopes and quantities are set up in a table of items to examine. As the required items regarding the packaging are determined, the table will be populated.

Isotope and Quantity	Form	Quantity A_1, A_2	Package A or B	Label, TI	Vehicle Placard
I-131, 10 mCi, unsealed in vial	**normal**	?	?	?	?
Cs-137, 8 mCi, sealed	**special**	?	?	?	?
Am-241, 20 Ci, sealed	**special**	?	?	?	?
Ir-192, 100 Ci, sealed	**special**	?	?	?	?

Table 4-1 Examples of Shipments—Form

The I-131 package doesn't meet the criteria for special form as it is contained in a vial which can be opened without destroying it. The Cs-137, Am-241, and Ir-192 sources qualify as special form as they are sealed sources.

Next, USDOT rule **§ 173.435, Table of A1 and A2 Values For Radionuclides** for $\mathbf{A_1}$ and $\mathbf{A_2}$ values are indicated in the table. For the 20 curie Am-241 source, the $\mathbf{A_1}$ value is 270 Ci (10 TBq) and, since 20 curies is less, we find that the $\mathbf{A_1}$ value applies. We repeat the process for the other isotopes. The unsealed I-131 meets the $\mathbf{A_2}$ criteria and the Cs-137 meets the $\mathbf{A_1}$ criteria. Finally, the Ir-192 source exceeds the $\mathbf{A_1}$ table value of 27

Ci (1 TBq) so Type B packaging is required. For the **A_1** or **A_2 packages**, we choose Type A shipping containers. For the Ir-192 source, a Type B Package will be required.

Isotope and Quantity	Form	Quantity A_1, A_2	Package A or B	Label, TI	Vehicle Placard
I-131, 10 mCi, unsealed in vial	normal	A_2	A	?	?
Cs-137, 8 mCi, sealed	special	A_1	A	?	?
Am-241, 20 Ci, sealed	special	A_1	A	?	?
Ir-192, 100 Ci, sealed	Special	$>A_1$	B	?	?

Table 4-2 Examples of Shipments—A_1 and A_2 Values, Package Types

Labels
Two (2) labels are required to be prominently displayed on opposite sides of each package. The three types of labels are:

Figure 4-1: White I, Yellow II, and Yellow III Package Labels.

The upper triangle of the Radioactive II and the Radioactive III marked labels (center and right, respectively) are yellow in color, hence, the names White I, Yellow II, and Yellow III.

Selection of the proper labeling is also based on the radiation levels around the package. A survey is performed at the package surface and at 1 meter from the surface. The results (maximum of each) are compared to the following table for determination of the label:

Transport index	Maximum radiation level at any point on the external surface	Label category[A]
0[B]	Less than or equal to 0.005 mSv/h (0.5 mrem/h)	WHITE-I.
More than 0 but not more than 1	Greater than 0.005 mSv/h (0.5 mrem/h) but less than or equal to 0.5 mSv/h (50 mrem/h)	YELLOW-II.
More than 1 but not more than 10	Greater than 0.5 mSv/h (50 mrem/h) but less than or equal to 2 mSv/h (200 mrem/h)	YELLOW-III.
More than 10	Greater than 2 mSv/h (200 mrem/h) but less than or equal to 10 mSv/h (1,000 mrem/h)	YELLOW-III (Must be shipped under exclusive use provisions)

Table 4-3 Category[36] of Label to be Applied to Class 7 (Radioactive) Materials Packages

[A] Any package containing a "highway route controlled quantity" (§173.403) must be labeled as RADIOACTIVE YELLOW-III.

[B] If the measured TI is not greater than 0.05, the value may be considered to be zero.

An item of information required to be entered on the label is the Transport Index (TI). USDOT defines the Transport Index[36] as: Transport Index (TI) means the dimensionless number (rounded up to the next tenth) placed on the label of a package, to designate the degree of control to be exercised by the carrier during transportation. The transport index is determined by multiplying the maximum radiation level in millisieverts (mSv) per hour at 1 m (3.3 ft) from the external surface of the package by 100 (equivalent to the maximum radiation level in millirem per hour at 1 m (3.3

[35] 49 CFR § 172.403(c)

[36] § 173.403 Definitions

ft)). The TI is the highest radiation level in mr/hr at 1 meter from the surface of the package. For example, a maximum measurement of 0.5 mr/hr at 1 meter would yield a TI of .5. Notice that the White I label will not require a TI since there is no radiation level at 1 meter to be considered. In addition, the maximum TI for a Yellow II is 1 while the Yellow III TI can exceed 1. The total TI (for all packages) can only exceed 50 when transport is in an exclusive use closed vehicle.

By surveying the (shielding containers) of the three sources, the radiation levels are determined to be:

I-131	15 mr/hr @ surface	0.4 mr/hr @ 1 meter	then the TI is 0.4
Cs-137	16 mr/hr @ surface	0.5 mr/hr @ 1 meter	then the TI is 0 .5
Am-241	40 mr/hr @ surface	1.8 mr/hr @ 1 meter	then the TI is 1.8
Ir-192	65 mr/hr @ surface	2.6 mr/hr @ 1 meter	then the TI is 2.6

Note: These values are fictitious and selected for illustration ONLY! ***Do not use them for actual transport procedures.***

From the survey information, the labels with the proper TIs are selected:

Isotope and Quantity	**Form**	**Quantity A_1, A_2**	**Package A Or B**	**Label, TI**	**Vehicle Placard**
I-131, 10 mCi, unsealed in vial	normal	A_2	A	**Yellow II 0.4**	?
Cs-137, 8 mCi, sealed	special	A_1	A	**Yellow II 0.5**	?
Am-241, 20 Ci, sealed	special	A_1	A	**Yellow II 1.2**	?
Ir-192, 100 Ci, sealed	Special	$>A_1$	B	**Yellow III 2.6**	?

Table 4-4 Examples of Shipments

Notes:

1. 1 mCi = $3.7E^7$ Bq or 37 MBq
2. The placard column will be populated later.

For the Yellow II and Yellow III labels, the contents (isotope), activity (in SI units), and TI need to be entered. The labels and entries would be:

I-131 package:

Label **(Yellow II)**
Contents: **I-131**
Activity: **370 MBq (10 mCi)**
TI: **0.4**

Cs-137 package:

Label **(Yellow II)**
Contents: **Cs-137**

Activity: **296 MBq (8 mCi)**

TI: 0.5

Yellow II
Enter information on Yellow II or III package label.

Am-241 package:

Label **(Yellow III)**
Contents: **I-131**
Activity: **740 GBq (20 Ci)**
TI: **1.8**

Iridium-192 package:

Label **(Yellow III)**
Contents: **Ir-192**
Activity: **3.7 TBq (100 Ci)**
TI: **2.6**

Shipping Paper

For each shipment/transport, a shipping paper, or bill of lading, must be prepared to accompany the shipment. The shipper is required to accurately prepare the document for all shipments, whether transport is performed by a licensee or by a common or contract carrier.

> *Note: the following discussion does not address hazardous waste shipments, which also involve EPA requirements. The information is not all-inclusive and is paraphrased or edited. Refer to 49 CFR Part 172 Subpart C, (172.200-205) for full information.*

General Requirements for Shipping Papers

DOT regulations[37] require the following information, as a minimum, be included on the shipping paper. It may also include additional information that the shipper feels in necessary for assuring that the shipment reaches its destination without accident or injury.

Shipping paper requirements include (refer to 49 CFR §172.101 Hazardous Materials Table for more information):

- A description of the hazardous materials is required. When including non-hazardous material on the same shipping paper, the hazardous material information must:
    1. Be entered first, or
    2. Be entered in a color that clearly contrasts with any description of non-hazardous material on the shipping paper; or
    3. Be identified by the entry of an "X" placed before the proper shipping name in a column captioned "HM". If appropriate, the "X" may be replaced by "RQ".
- The required shipping description on a shipping paper and all copies must be legible and printed, either manually or mechanically in English.
- The required shipping description may not contain any code or abbreviation—unless specifically authorized or required in the regulations.
- Optional information—The shipping paper may contain additional information concerning the material provided the information is not inconsistent with the required description. Unless otherwise permitted or required, additional information must be placed after the basic required description.
- Electronic shipping papers are permitted for transportation by rail. See CFR 49 § 172.201(a)(5) for special requirements.

---

37 49 CFR § 172.201 Preparation and retention of shipping papers (1/9/2013).

- Continuation pages may be used if each page is numbered consecutively and the first page indicates the total number of pages—such as Page 1 of 6 pages.
- An emergency response phone number must be included.

Retention and Recordkeeping

Each shipping paper must:

- Have a copy (or an electronic image) retained at the principle place of business.
- Be made available to Federal, state, and local government agencies at reasonable times and locations.
- Be maintained for two (2) years for hazardous materials, or three (3) years for hazardous waste.
- Include the date of acceptance by the initial carrier.

If a licensee carries its own hazardous material, such as radiography, well-logging, or moisture density gauge companies do, then the following rule is useful:

> *"A motor carrier (as defined in §390.5 of subchapter B of chapter III of subtitle B) using a shipping paper without change for multiple shipments of one or more hazardous materials having the same shipping name and identification number may retain a single copy of the shipping paper, instead of a copy for each shipment made, if the carrier also retains a record of each shipment made, to include shipping name, identification number, quantity transported, and date of shipment."*

Thus, a licensee that uses the same shipping paper for multiple shipments can maintain a copy of that record and copies of the shipping or transport logs to satisfy the shipping record retention requirement, provided the log contains:

- each shipment made,
- shipping name,
- identification number,
- quantity transported, and
- date of shipment

Specific Requirements of Shipping Paper Entries

Much of the following required information can be obtained from the manufacturer/supplier. They usually ship the radioactive material to the licensee, so their shipments and packages will most likely meet the current requirements. However, it is the responsibility of the shipper (which will be the licensed user) to meet all of the requirements.

Description Requirements

The description of hazardous material must include (refer to table of 49 CFR §172.101):

1. identification number;
2. proper shipping name;
3. hazard class or division;
4. packing group in Roman numerals (NOT required for radioactive material); and
5. the total quantity (activity) of hazardous materials covered by the description

Items 1, 2, 3, and 4 must be shown in sequence with no additional information interspersed. For example, a portable moisture density gauge containing only 8 mCi of Cs-137 would be shown as "**UN3332, Radioactive Material, Type A Package, Special Form, 7**." The entries listed below do not need to be in any particular order.

Reportable Quantity (RQ)

According to US Environmental Protection Agency (EPA) rules, *Reportable Quantity* ("RQ") means "*that quantity, as set forth in this part*[38]*, the release of which requires notification pursuant to this part*". Reports would be made to the NRC or state authority depending on certain circumstances. Most licensees will not be affected by this requirement. The reportable quantities can be found in the EPA table: TABLE 302.4—LIST OF HAZARDOUS SUBSTANCES AND REPORTABLE QUANTITIES, found in 40 CFR § 302.4.

Additional Required Entries for Descriptions

For descriptions of radioactive materials (Class 7), the following must also be included:

1. The **name of each radionuclide** (refer to table § 173.435 Table of A1 and A2 values for radionuclides.).
2. A description of the physical and chemical form of the material, if the material is not in special form (generic chemical description is acceptable for chemical form).
3. The activity contained in each package of the shipment in terms of the appropriate SI units (e.g. Becquerel, Terabecquerel, etc.) or in terms of the appropriate SI units followed by the customary units (e.g. Curies, millicuries, etc.).
4. The category of label applied to each package in the shipment (example: RADIOACTIVE WHITE-I).
5. The transport index assigned to each package in the shipment bearing RADIOACTIVE YELLOW-II or RADIOACTIVE YELLOW-III labels.
6. For a package approved by the. US DOE or US NRC, a notation of the package identification marking.

---

38 40 CFR Part 302—Designation, Reportable Quantities, And Notification

7. For an export shipment or a shipment in a foreign made package, a notation of the package identification marking as prescribed in the applicable International Atomic Energy Agency (IAEA) Certificate of Competent Authority.
8. For a shipment required as exclusive use:
   - An indication that the shipment is consigned as exclusive use; or
   - If all the descriptions on the shipping paper are consigned as exclusive use, then the statement "Exclusive Use Shipment" may be entered only once on the shipping paper in a clearly visible location.
9. An **emergency response telephone number**.

The shipping paper must be provided to the driver and placed in a position within immediate reach of the driver during transport.

Identification Number

An identification number is assigned to each proper shipping name. If the number is preceded by the letters "UN", they are associated with proper shipping names that are considered appropriate for international transportation as well as domestic transportation. The following table lists the identification numbers for Class 7 (radioactive) materials. Please note that this information could be changed by US DOT at any time. The identification numbers for radioactive materials are shown in Table 4-5.

Table 4-5 Descriptions and Proper Shipping Names, Class and Identification Numbers for Radioactive Materials
(Extracted from 49 CFR §172.101 Hazardous Materials, Jan 2013)

Hazardous Materials Descriptions and Proper Shipping Names	Hazard Class or Division	Identification Numbers
Radioactive material, excepted package-articles manufactured from natural uranium or depleted uranium or natural thorium	7	UN2909
Radioactive material, excepted package-empty packaging	7	UN2908
Radioactive material, excepted package-instruments or articles	7	UN2911

Radioactive material, excepted package-limited quantity of material	7	UN2910
Radioactive material, low specific activity (LSA-I) non fissile or fissile excepted	7	UN2912
Radioactive material, low specific activity (LSA-II) non fissile or fissile excepted	7	UN3321
Radioactive material, low specific activity (LSA-III) non fissile or fissile excepted	7	UN3322
Radioactive material, surface contaminated objects (SCO-I or SCO-II) non fissile or fissile excepted	7	UN2913
Radioactive material, transported under special arrangement, non fissile or fissile excepted	7	UN2919
Radioactive material, transported under special arrangement, fissile	7	UN3331
Radioactive material, Type A package, fissile non-special form	7	UN3327
Radioactive material, Type A package non-special form, non fissile or fissile-excepted	7	UN2915
Radioactive material, Type A package, special form non fissile or fissile-excepted	7	UN3332
Radioactive material, Type A package, special form, fissile	7	UN3333
Radioactive material, Type B(M) package, fissile	7	UN3329
Radioactive material, Type B(M) package non fissile or fissile excepted	7	UN2917
Radioactive material, Type B(U) package, fissile	7	UN3328
Radioactive material, Type B(U) package non fissile or fissile excepted	7	UN2916
Radioactive material, uranium hexafluoride non fissile or fissile excepted	7	UN2978
Radioactive material, uranium hexafluoride, fissile	7	UN2977

Transporting Radioactive Material

After the proper packaging and labeling have been completed and the shipping paper has been prepared, the appropriate vehicle must be selected and, when necessary, placarded to warn of the radiation hazard.

Transport Requirements

Vehicles, Placards

For purposes of transport, there are three types of carriers: common, contract, and private. Common and contract carriers provide a service to others by carrying their materials. Contract carriers establish a contract for transport services but common carriers publish their rates. They are not licensed by an agency for transport of radioactive material. Private carriers are licensees that transport radioactive materials for their own business. The type of transport/vehicle determines which radiation level limits apply.

Type of Transport	Package Radiation Limits	Vehicle Radiation Limits
Common Carrier, Non-Exclusive Use		
Closed	200 mrem/hr (2 mSv/hr) on the surface of the package and 10 mrem/hr (0.1 mSv/hr) at 1 meter from any surface of the package.	N/A
Open	Same as above.	N/A
Contract Carrier, Exclusive Use		
Closed	1000 mrem/hr (10 mSv/hr) on the surface of the package.	200 mrem/hr (2 mSv/hr) at the surface of the vehicle, 10 mrem/hr (0.1 mSv/hr) at 2 meters from any surface of the vehicle, and 2 mrem/hr (0.02 mSv/hr) in the vehicle cab.

Open	200 mrem/hr (2 mSv/hr) on the surface of the package.	200 mrem/hr (2 mSv/hr) on any imaginary surface of the vehicle, 10 mrem/hr (0.1 mSv/hr) at 2 meters from any imaginary surface of the vehicle, and 2 mrem/hr (0.02 mSv/hr) in the cab of the vehicle.
Table 4-6: Acceptable Radiation Levels for Types of Transport (From: Transportation of Radioactive Material, USNRC Technical Training Center, Rev 0703, P. 11-19)		

It should be noted that, for common carriers that carry both non-radioactive and radioactive packages simultaneously, the package radiation limit is used for control. While the package limits are higher for contract and private carriers carrying only radioactive material, the radiation limits around the vehicle are used for control. However, the packages that we are setting up for transport will be carried by private carrier in an exclusive use vehicle—closed transport.

When loading the vehicle, the package(s) must be blocked and braced to prevent movement or shifting during transport. This is a specific requirement[39].

Placard

Certain shipments of packages containing radioactive material require that a readily visible placard be placed on each of the four (4) sides of the vehicle. The placard is diamond shaped, like the shipping labels, but is much larger. Vehicles are required to placarded if they carry packages displaying a Yellow III label, or transport LSA material, or transport "highway route controlled quantities". **For highway route controlled quantities, the placard is surrounded by a black square border**. However, the second two types of shipments are not addressed here.

39 49 CFR 177.842 (d)

Figure 4-2 Vehicle Placard

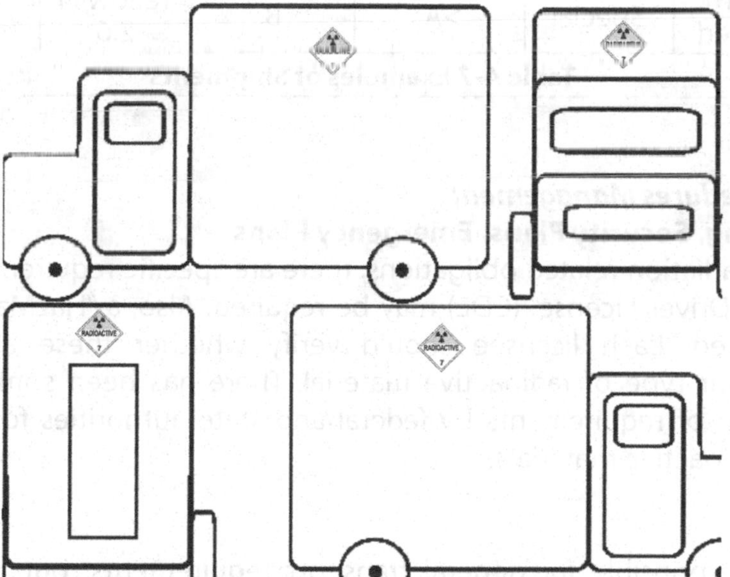

Figure 4-3 Placard Placement on Vehicle

The above drawing illustrates example placard placements. The placards can be placed anywhere on the side provided that they are clearly visible.

For the selected packages, then, only the Ir-192 package requires placarding of the vehicle. We can now complete the final column of the table.

Isotope and Quantity	Form	Quantity A_1, A_2	Package A Or B	Label, TI	Vehicle Placard
I-131, 10 mCi, unsealed in vial	normal	A_2	A	YellowII 0.4	**No**
Cs-137, 8 mCi, sealed	special	A_1	A	Yellow II 0.5	**No**
Am-241, 20 Ci, sealed	special	A_1	A	Yellow II 1.2	**No**
Ir-192, 100 Ci, sealed	Special	$>A_1$	B	Yellow III 2.6	**Yes**

Table 4-7 Examples of Shipments

Transport Procedures Management

Drivers, Training, Security Plans, Emergency Plans

In addition to radiation related obligations, there are specific requirements for drivers. A "Commercial Driver License" (CDL) may be required. Also, a "HazMat" endorsement may be required. Each licensee should verify whether these are required for transport of their type of radioactive material. There has been some disagreement in interpretation of requirements by federal and state authorities for some types of transport of radioactive materials.

Driver

The driver is responsible for general transport requirements, but in particular, for maintaining the radioactive package transport requirements during the process. The driver, as a minimum, should assure:

- the package(s) remain properly blocked and braced until reaching the destination;
- the package labels remain legible;
- the placards (if required) are properly placed on the transport vehicle;
- his/her CDL (if required) is current;
- the shipping paper is within easy reach at all times during transport and that the emergency instructions are clear and understood; and

- during stops for meals and personal needs, the vehicle is parked in a location where it can be readily observed and is away from areas frequented by people—if possible.

The shipper must provide the driver with special instructions when shipping as "exclusive use". The driver should review the shipping paper before departure and discuss any unclear information with the shipper.

Hazardous Materials Transport Training Requirements

Who must be trained? US DOT rules require that each hazardous materials employee involved in the transport of hazardous materials, which includes radioactive materials, be provided initial and refresher training pertinent to his/her particular job/duties.

Note: The following is excerpted/paraphrased from CFR 49, Subpart H, §172.700-704.

According to 49 CFR § 171.8:
Hazmat employee means:

(1) *A person who is:*
  - (i) *Employed on a full-time, part time, or temporary basis by a hazmat employer and who in the course of such full time, part time or temporary employment directly affects hazardous materials transportation safety;*
  - (ii) *Self-employed (including an owner-operator of a motor vehicle, vessel, or aircraft) transporting hazardous materials in commerce who in the course of such self-employment directly affects hazardous materials transportation safety; (iii) A railroad signalman; or (iv) A railroad maintenance-of-way employee.*

(2) *This term includes an individual, employed on a full time, part time, or temporary basis by a hazmat employer, or who is self-employed, who during the course of employment:*
  - (i) *Loads, unloads, or handles hazardous materials;*
  - (ii) *Designs, manufactures, fabricates, inspects, marks, maintains, reconditions, repairs, or tests a package, container or packaging component that is represented, marked, certified, or sold as qualified for use in transporting hazardous material in commerce.*
  - (iii) *Prepares hazardous materials for transportation;*
  - (iv) *Is responsible for safety of transporting hazardous materials;*
  - (v) *Operates a vehicle used to transport hazardous materials.*

and

Hazmat employer means:

(1) *A person who employs or uses at least one hazmat employee on a fulltime, part time, or temporary basis; and who:*
   - *(i) Transports hazardous materials in commerce;*
   - *(ii) Causes hazardous materials to be transported in commerce; or*
   - *(iii) Designs, manufactures, fabricates, inspects, marks, maintains, reconditions, repairs or tests a package, container, or packaging component that is represented, marked, certified, or sold by that person as qualified for use in transporting hazardous materials in commerce;*

(2) *A person who is self-employed (including an owner-operator of a motor vehicle, vessel, or aircraft) transporting materials in commerce; and who:*
   - *(i) Transports hazardous materials in commerce;*
   - *(ii) Causes hazardous materials to be transported in commerce; or*
   - *(iii) Designs, manufactures, fabricates, inspects, marks, maintains, reconditions, repairs or tests a package, container, or packaging component that is represented, marked, certified, or sold by that person as qualified for use in transporting hazardous materials in commerce; or*

(3) *A department, agency, or instrumentality of the United States Government, or an authority of a State, political subdivision of a State, or an Indian tribe; and who:*
   - *(i) Transports hazardous materials in commerce;*
   - *(ii) Causes hazardous materials to be transported in commerce; or*
   - *(iii) Designs, manufactures, fabricates, inspects, marks, maintains, reconditions, repairs or tests a package, container, or packaging component that is represented, marked, certified, or sold by that person as qualified for use in transporting hazardous materials in commerce.*

Thus, if your company is involved in hazardous materials transport in any form or fashion, and if any of your employees are involved in operations of transport of these materials, then those employees are hazmat employees and all of the training and operational requirements for hazmat transport must be met. During a meeting once attended by the author, a US DOT person gave the following example to illustrate how to determine the status of employees: "If a company secretary prepares a hazmat shipping paper from a document prepared by a trained hazmat person, then the secretary is not a hazmat employee. If, however, the secretary looks up, say, a package identification number, then he/she would be a hazmat employee and the training would be required."

Training

According to US DOT definitions, training means *"a systematic program that ensures a hazmat employee has familiarity with the general provisions of this subchapter (49 CFR*

172), is able to recognize and identify hazardous materials, has knowledge of specific requirements of this subchapter applicable to functions performed by the employee, and has knowledge of emergency response information, self-protection measures and accident prevention methods and procedures".

One should keep in mind that the training is required for all classes of hazmat, not simply radioactive materials (Class 7). Additional training (modal-specific) requirements for individual modes of transportation are prescribed in 49 CFR § 174, 175, 176, and 177, but they are not addressed here. Also, a state may have more stringent training requirements if they are not in conflict with the training requirements of 49 CFR § 172 and 177 and they only apply to drivers domiciled in that state.

Initial and Refresher Training

Two types of training are required: initial and refresher. Initial training must be provided within 90 days of beginning the job function (or changes in duties). Refresher (or recurring) training must be performed every three (3) years. Training by a previous employer or source can be used to satisfy the requirement if the training was appropriate and a record is available. A new employee, or an employee that changes job duties, can perform hazardous materials related duties before the training is completed provided that the individual performs the duties under the direct supervision of a "properly trained and knowledgeable hazardous materials employee" and the employee receives the required training within "90 days of employment or change in job function".

Training Content

The initial training program must address:

- General awareness/familiarization training: training designed to provide familiarity with the transport requirements and enable the employee to recognize and identify hazardous materials.
- Function-specific training: training concerning US DOT transport requirements, including any issued exemptions or special permits, that are specific to the functions the employee performs (an alternative is to provide training relating to the requirements of the ICAO Technical Instructions and/or the IMDG Code if such training addresses functions authorized by 49 CFR § 171).
- Safety training: training concerning emergency response information, measures to protect the employee from the hazards associated with hazardous materials to which they may be exposed in the work place, including specific measures the hazmat employer has implemented to protect employees from exposure, methods

and procedures for avoiding accidents, such as the proper procedures for handling packages containing hazardous materials.

- Security awareness training: training that provides an awareness of security risks associated with hazardous materials transportation and methods designed to enhance transportation security, including a component covering how to recognize and respond to possible security threats. New hazmat employees must receive the security awareness training within 90 days after employment.
- In-depth security training: training that provides each hazmat employee of a company/person required to have a security plan required by 49 CFR § 172.800 who handles hazardous materials covered by the plan, performs a regulated function related to the hazardous materials covered by the plan, or is responsible for implementing the plan must be trained concerning the security plan and its implementation. The security training must include:
    - company security objectives;
    - organizational security structure;
    - specific security procedures;
    - specific security duties and responsibilities for each employee; and
    - specific actions to be taken by each employee in the event of a security breach.

Hazard communication programs training required by other federal agencies, such as OSHA and EPA, and by international agencies, or training conducted by employers to comply with security training programs required by other Federal or international agencies, may be used to satisfy the training requirements if such training addresses the above training components.

Training Records

Training records must include "the hazardous materials employee's name; completion date of most recent training; training materials (copy, description, or location); name and address of the hazardous materials trainer; and certification that the hazardous materials employee has been trained and tested."

The employer is responsible for providing the training and maintaining the documentation. The employer may provide the training through the company's own resources, or may retain an independent, qualified trainer. If an outside trainer is used, then records showing the trainer's qualifications should also be maintained.

Resources for transport emergencies must be made available to the driver during transport. Emergency procedures must be provided with the shipping paper. They do not need to be elaborate—but should indicate immediate action to be taken if an accident occurs.

Safety and Security Plans

Companies/persons that offer for transportation in commerce or transports in commerce one or more of the hazardous materials listed in 49 CFR § 172.800 must develop and adhere to a transportation security plan for hazardous materials that conforms to the requirements of Subpart I—Safety and Security Plans, of 49 CFR § 172. The Class 7 materials listed are:

- A quantity of uranium hexafluoride requiring placarding under §172.505(b);
- International Atomic Energy Agency (IAEA) Code of Conduct Category 1 and 2 materials, including Highway Route Controlled quantities as defined in 49 CFR 173.403 or known radionuclides in forms listed as RAM-QC by the Nuclear Regulatory Commission.

Most licensees, outside of industrial radiography and well-logging, will not have to incorporate security plans. If your company transports these materials, consultation with persons expert in this area is recommended. Specific details/requirements for security plans can be found in CFR § 172.802 Components Of A Security Plan.

Transport Emergencies

A general approach is to include steps for:

- Securing the area of the accident.
- Calling for assistance (list of emergency phone numbers, including state and federal emergency assistance resources—911 can be used for local emergency assistance in most areas).
- Maintaining control over the sources and/or package(s), and area, until help arrives.

4. Responding to Emergencies

Radiation Emergencies

Notification, Reporting

As in any other of man's hazardous endeavors, uses of radiation often lead to accident and/or injury. Some are merely a nuisance while others can produce serious injury—even death. There is a broad spectrum of possibilities. When working with radiation sources, a good practice is to assume that an accident will occur—so we should perform each step of the operation carefully and allow for rapid recovery from any accident. Radiation control regulations generally refer to accidents as "radiation incidents" or "radiation emergencies". Radiation incidents meeting certain criteria ARE REQUIRED to be reported within specific time frames. If you timely report an incident, you will probably receive (free) agency assistance/guidance and avoid higher monetary penalties. If you fail to timely report, you will probably face increased penalties.

> Note: *The information in this section was obtained from TAC §289.202(xx), and should be nearly identical in all federal/state radiation control rules.*

Immediate Notification

Incidents involving radiation sources and meeting the following criteria are required to be reported to the appropriate agency **IMMEDIATELY**:

- An incident which causes, or threatens to cause, a person to receive—
  - a total effective dose equivalent (TEDE) of 25 rems (0.25 Sv) or more;
  - an lens dose equivalent of 75 rems (0.75 Sv) or more; or
  - a shallow dose equivalent to the skin or extremities, or a total organ dose equivalent, of 250 rads (2.5 grays) or more;
- An incident which causes, or threatens to cause the release of radioactive material, inside or outside of a restricted area, so that, had an individual been present for 24 hours, the individual could have received an intake five times the occupational ALI. (Does not apply to routine work stations—which are covered under other provisions.)

Twenty-four (24) Hour Notification

Incidents/events involving the loss of control of a licensed or registered radiation source and meeting the following criteria are required to be reported to the appropriate agency within 24 hours of discovery of the event:

- An incident/event which may have caused, or threatens to cause an individual to receive, in a period of 24 hours;
  - a total effective dose equivalent (TEDE) exceeding 5 rems (0.05 Sv);
  - an eye dose equivalent exceeding 15 rems (0.15 Sv);
  - a shallow dose equivalent to the skin or extremities or a total organ dose equivalent (TODE) exceeding 50 rems (0.5 Sv).
- An incident/event which causes the release of radioactive material, inside or outside of a restricted area, so that, had an individual been present for 24 hours, the individual could have received an intake in excess of one occupational ALI. (Does not apply to routine work stations—which are covered under other provisions.)

Preparation for Radiation Emergencies

Accidents, Incidents, and Emergencies

Proper preparation for the types of possible emergencies will prepare a help to quickly recover from an accident with a minimum of injury and damage. In preparation for emergencies, one should identify available resources. The resources within a company are generally known to management. Resources outside of a company include government agencies, other industries, medical facilities, and educational facilities.

Preparation: Identify Resources, Prepare Written Procedures

Identify Resources

When an accident occurs within a facility, most likely only the governmental radiation control agency and local emergency response agencies will be involved. There is, of course, a requirement to notify the radiation control agency (as discussed above) of an accident and such an agency will often work with the licensee and with local officials to respond appropriately. Local authorities can assist with fire, theft, and injuries of people, while the state or federal agencies can provide health physics staff to help identify hazards and determine effective controls. They also provide guidance in accident recovery and generally must approve methods and procedures for recovery—particularly if radioactive contamination is present. Local, county and state police can assist with locating stolen vehicles or radiation sources. Prior arrangements should be made with a hospital equipped to deal with radiation injuries and contaminated persons to provide for medical needs. The radiation control agency can often assist in locating nearby, and national, medical resources. Many educational facilities have research labs where radioactivity and radiation sources are used—so they also have radiation survey and measurement instruments and staff familiar with radiation problems. There are a number of industrial companies that provide services relating to radiation sources. These should be identified and considered in preparing emergency plans.

Table 4-8: Potential Emergencies to Consider for Response Preparation

General emergency procedure	Emergency due to radioactive contamination
General field emergency procedure	Emergency due to vehicle accident
Theft of a source/device	Emergency causing exposure of unmonitored persons
Loss of a source/device	Emergency involving fire

Emergency due to source/device equipment malfunction	Emergency source recovery
Emergency due to release of radioactive material	Emergency due to x-ray machine equipment malfunction

Prepare Written Procedures
The programs that successfully and rapidly recover, even prevent, radiation emergencies and accidents will be the ones that have taken the time to properly design and implement operating procedures, and also consider all possible operating conditions, and will have prepared suitable written emergency procedures and have trained employees in using them. This is true in any hazardous operation—radiation related or other. Although emergencies that may actually occur will probably be different than those planned for, personnel will be able to make adjustments for the differences and react more properly than personnel who have little or no training.

Radiation Safety Officer Responsibility: For radiation emergencies, in general, the radiation safety officer is responsible for timely notifying all appropriate authorities, assuring the site of the emergency and the radiation source(s) are secured and persons are protected from radiation hazards, performing necessary leak tests, and arranging for proper shipment or disposal of the source or device, if necessary. The radiation safety officer is also responsible for documenting the accident and providing the appropriate reports to the agency(s) with jurisdiction. The services of a qualified consultant in radiation safety may also be used to assist in any of the procedures.

General Emergency Procedure
Radiation Worker, Radiation Safety Officer
A general approach to handling a radiation emergency, particularly one away from the licensee's facilitates, should consider, as a minimum:

For the Radiation Worker:
A worker encountering a radiation problem, in most areas of use, can use the following steps and generally protect him/herself and others from unnecessary, even harmful, radiation exposure:

- *Secure the area:* Secure the area of the emergency by establishing a visible barrier at a minimum radius of 50 feet from the center point of the source location (a much greater distance will probably be required for unsealed radioactive material and an even greater distance int the down-wind direction).

- *Contact the radiation safety officer:* The worker should immediately contact the radiation safety officer by telephone, radio, etc., and report the circumstances of the emergency.
- *Inspect damage*: The container(s) should be inspected, as advised by the radiation safety officer for damage. The radiation safety officer may have to travel to the site and inspect the damage. If a radiation survey instrument is available, the worker should perform a survey to establish an exclusion zone and safe working areas. If a survey instrument is not available, pre-established zones should be initiated.
    - *Minor damage:* If the damage appears to be minor and the source is safely and the radiation safety officer authorizes, the source or device may be placed in its transport container and returned to the storage area for an evaluation and emergency leak test.
    - *Serious damage:* If the source appears to be severely damaged, i.e. broken, crushed or burned, a barrier around the source should be established (if possible, using warning tape, rope, signs) and all persons should be barred from the area. The radiation worker should remain physically present at all times to prohibit entry into the restricted area by unauthorized/unmonitored persons.
- *Record information*: If the radiation safety officer is not immediately present at the site of the accident, the radiation worker should record all of his/her observations regarding the conditions of the emergency, including, but not limited to, the following:
    - The date/time and conditions at the time of the accident; and
    - the names, addresses, and phone numbers of persons present (within the vicinity of the location of the source involved).

    Note: Persons in the vicinity need to be identified so that their potential exposure can later be determined or estimated.

For the Radiation Safety Officer:

- *Contact authorities:* Timely contact appropriate authorities—their resources may be needed.
- *Assure Area Secured:* Assure the site of the emergency and the source are secured and persons are protected from radiation hazards (coordinate with radiation worker on the scene).
- *Perform Leak Tests:* Perform necessary leak tests.
- *Arrange for Transport/Disposal:* Arrange for proper shipment or disposal of the source or device, if necessary.
- *Prepare Documentation and Reports:* Document the incident and provide the appropriate reports to the agency with jurisdiction.

Clearly, a licensee can enhance the chance of a successful recovery from an accident by preparing written procedures, and training personnel to know and understand them, which address the most likely types of accidents that might be encountered in the type of operation(s) used.

It would not be possible to discuss all of the many different types of radiation emergencies and their many variations in this one book. The above radiation emergency "situations" may be encountered and examples and suggested protective steps are presented in **Appendix 4**. They are not all inclusive. Each program must evaluate its own circumstance and prepare to respond to radiation emergencies, accordingly.

Chapter 5

Regulatory Requirements and Processes

1. Introduction

There are many state and federal agencies involved in regulating the various uses of radiation. This chapter discusses how the various agencies fit into "the scheme of things". Since there are 50 states, perhaps 6-10 federal agencies, and several cities and counties that potentially can be involved, it would be very difficult to visit all possibilities. It is better to try to comprehend the overall regulatory philosophy for radiation control processes and adapt to the individual differences—and periodic changes.

2. Requirements of Federal and State Regulations

The Regulatory Process

Development of Radiation Control, Agreement State and Federal Programs, The Texas Radiation Control Program, Summary of Regulatory Agencies

Radiation Control Authorities

Today, there are many *radiation control authorities* in place to protect the public from radiation hazards. During the growth and development of the use of radiation sources, safety hazards for the worker and the public were recognized and so government developed regulatory programs to provide controls. Many federal and state government agencies, or subunits of agencies, were created and complex, interlocking authorities and jurisdictions developed. A brief overview of the development is provided to allow some understanding—or, at least an appreciation—of these complexities. It should be noted that many citizens turn to the federal government when difficulties develop, but it is often the task of state or local government to deal with the majority of day-to-day radiation related problems.

Development of Radiation Control

Early in the radiation use era, all radioactive material except radium (naturally occurring) was owned by the federal government. Radium, thought to be valuable and to increase in value over time, was purchased and stored by many individuals. Much of it was stored in safes or safety deposit boxes at banks. However, as the value dropped these caches of radium became liabilities as they offered high disposal costs and major contamination problems. Later, the federal government gave up ownership of radioactive material and simply established, through the "Atomic Energy Act", a method of licensing radioactive materials for potential users. Essentially, any radioactive material produced by, or through, nuclear reactors was required to be licensed by the (then) Atomic Energy Commission (AEC). Any other radioactive material (naturally occurring or machine produced) was left "unregulated". X-ray machine use was also unregulated. Many states, then, created radiation control programs (usually within the states' health departments) to assure the safe use and control of naturally occurring or machine produced radiation sources. The federal government eventually established authority over manufacturers of radiation producing machines and set standards for manufacturers to follow and, currently, the US Food and Drug Administration (FDA) maintains these controls, but does not regulate at the "user level". This is left to the states. The federal government also established controls over transport of hazardous materials in interstate commerce. Radioactive material is a hazardous material and so falls under the regulations developed by the US Department of Transportation (DOT). Most agreement states incorporate these rules, all or in part, by adopting them or incorporating them into their own regulations. In many cases, a state's highway patrol may have the DOT rules incorporated into their vehicle code—allowing dual or overlapping jurisdiction.

Agreement State and Federal Programs

In the late 1950's and early 60's, the AEC initiated a process of developing "agreements" with interested state radiation control programs to allow the states to license and regulate the use of radioactive materials within the state. These are called "Agreement States". The applicant state was required to establish a licensing program that was "compatible" with the federal radiation control program. The AEC retained regulation of nuclear power plants, federal operations, and licensing of companies manufacturing and distributing "exempt sources" of radioactive material. Successful states took on the process of licensing the former AEC licensees, and any new ones—and accepted all of the responsibilities associated with those licenses. The governor signed Texas' agreement in 1963 and the program was assigned to the Department of Health (now DSHS) where it was situated in the Division of Occupational Health and Radiation Control. This was a typical arrangement for many new Agreement States. The DSHS then developed the Texas Regulations for Control of Radiation to enable regulation of all radiation use under the state's jurisdiction. In the early 1980's, the Nuclear Regulatory Commission (formerly the AEC) also required states that licensed uranium mines to apply for and sign a separate agreement—which Texas did. The setting of radiation protection standards is an odd process. The International Commission of Radiation Protection (ICRP) develops general protection standards. The US's National Commission of Radiation Protection (NCRP) uses the ICRP recommendations to develop national recommendations for the US. The NRC, DOE, EPA, and DOT, as a minimum, then develop regulations to address the NCRP recommendations. Agreement States must then use the standards promulgated by the NRC, and, in some areas the EPA, to develop compatible regulations. Once, implemented by the state, they generally apply to all users—even x-ray machine users not regulated by the NRC or EPA.

The Conference of Radiation Control Program Directors (CRCPD) is an organization whose primary membership is composed of radiation control program directors from each state. It is a "nonprofit non-governmental professional organization dedicated to radiation protection" with the mission "to promote consistency in addressing and resolving radiation protection issues, to encourage high standards of quality in radiation protection programs, and to provide leadership in radiation safety and education[40]". States with 2 or more radiation control programs, such as one from the health department and one from the environmental protection organization, may have 2 or more members. The organization is funded by the states and by federal organizations, such as the NRC, EPA, DOT, etc. The state members are voting members while the federal organizations provide resource persons. The Suggested State Regulations for Control of Radiation (SSRCR or SSR) are developed by the CRCPD to provide states with a "template" for generating and implementing regulations that are compatible with those of the NRC and other states. The SSR can be accessed online through the CRCPD's website and can be useful for understanding regulatory requirements.

The Texas Radiation Control Program

Knowing how radiation control programs are organized helps one to better understand the regulatory process. Using the Texas radiation control program (RCP) as an example, we note that today, Texas has probably one of the most extensive radiation control programs in the world. The major portion of the current program is assigned (by law) to the Department of State Health Services (DSHS, formerly

40 http://www.crcpd.org/about/about.aspx

the Texas Department of Health, or TDH). The program was managed by Bureau of Radiation Control (BRC) program from 1981 until the early 2004. It is now the Division for Regulatory Services, Radiation Control Program and the duties are assigned to four (4) units: Licensing; Inspections; Policy, Standards and Quality Assurance (PSQA); and Enforcement. Licensing receives applications for licensing or registration of radiation sources and issues the permits. Inspections manages the inspection program for all radiation permits, performs investigation of accidents and incidents, and performs environmental surveillance. It also manages the radiological portions of the state's Emergency Response Plan (which covers all types of major emergencies—tornado, hurricane, flood, etc.). PSQA manages development and modification of radiation control regulations and processes inspection reports sent from Inspections. Enforcement handles administrative penalties and civil and criminal cases involving radiation sources. Enforcement also conducts similar actions for other health laws. Disposal of radioactive waste (not generation, processing, or storage) is assigned by law to the Texas Commission on Environmental Quality (TCEQ, formerly the Texas Natural Resources Conservation Commission). This agency has been processing license applications which would authorize the development and use of a radioactive waste site for the state. The TCEQ's authority also extends to disposal of naturally occurring radioactive material (NORM) waste—except for NORM waste that is generated through oil and/or gas processes. Disposal of this type of waste is regulated by the Texas Railroad Commission (TRRC), by law. The TRRC regulates oil and gas related activities in Texas. The DSHS regulates NORM uses, storage, processing, waste processing and storage, etc., while the TCEQ and TRRC regulate NORM disposal, only. On the federal level, the US Department of Energy (DOE) and the US Department of Defense (DOD) maintain authority over nuclear weapons. On the local government level, city, municipality, and county governments are generally preempted—with a few exceptions. They may have zoning ordinances, however, that influence or affect radiation uses. The US Environmental Protection Agency (EPA) has some authority (regarding releases of radioactive material to air/water) but it has been slow in developing standards for all areas. For transportation of radioactive materials, the Texas Department of Public Safety (DPS) and the Texas Department of Transportation (TXDOT) have varying authorities. DSHS, DPS, and TXDOT work jointly for setting transport routes. Otherwise, DSHS and DPS manage the general transport activities. DSHS performs inspections at licensee's sites and DPS performs vehicle inspections at check points.

It is very important for licensees and registrants to become familiar with the regulatory authorities that have jurisdiction over their operations. Most Agreement States will have agencies that operate similar to the Texas agencies discussed above. Some functions may be combined, or even more greatly spread through various agencies in the state. Radiation control programs can be placed in state health departments, environmental agencies, and/or labor departments (or any combination thereof). Two states (California and New York) have split programs. In NRC states, radioactive material is regulated by the NRC and x-ray devices are usually regulated by the state health department. The NRC has recently added regulation of certain NORM to its area of authority. It should be noted that OSHA has no authority in radiation matters in states which have radiation control programs.

Summary of Regulatory Agencies

The following list summarizes the pertinent radiation functions, agencies and most of the authorities discussed above (caution—these can change at any time) affecting Texas. For other Agreement States, substituting the particular state's radiation control program for Texas's will give a similar list. For non-agreement states, the NRC retains jurisdiction for nuclear reactor produced radioactive materials. Table 5-1 shows various state and federal agencies that have varying jurisdiction over radiation sources in Texas.

Table 5-1 Federal and State Agencies with Radiation Control Functions in Texas

Function	Texas/State	Federal
Radioactive materials, use, storage, processing, and all waste except disposal	DSHS	NRC: nuclear reactors and federal licensees
Waste disposal (except oil/gas NORM)	TCEQ	NRC, EPA partly
NORM disposal—oil/gas	TRRC	EPA partly
NORM use, storage, processing	DSHS	EPA partly
Radioactive materials routing	DSHS, DPS, TXDOT	DOT
Radioactive materials/ waste transport	DSHS (state licensees), DPS (all)	DOT, NRC: nuclear reactors and federal licensees
Radiation producing machines use	DSHS	None

Radiation producing machines manufacture, distribution	DSHS, partly	FDA
Laser device use	DSHS	FDA, FAA for laser light shows that may affect aircraft
Laser device manufacture, distribution	DSHS, if registered	FDA
Use of medical devices incorporating radiation sources	DSHS	None
Manufacture, distribution of medical devices incorporating radiation sources	DSHS (for certain device approvals)	FDA

DSHS = Texas Department of State Health Services (formerly Texas Department of Health [TDH])
TCEQ = Texas Commission on Environmental Quality
TRRC = Texas Railroad Commission
DPS = Texas Department of Public Safety
TXDOT = Texas Department of Transportation
NRC = United States Nuclear Regulatory Commission
EPA = United States Environmental Protection Agency
DOT = United States Department of Transportation
FDA = United States Food and Drug Administration
FAA = Federal Aviation Administration

Licenses and Registrations for Radiation Sources

Radioactive Material Licenses: Specific and General

Understanding which sources need to be licensed and which registered confuses many people and companies. In general, a license is required for the use/possession of radioactive material and a registration is required for a machine produced radiation source (x-ray or laser). There are overlaps, such as an accelerator using both machine produced radiation and radioactive material. A license must be possessed by the user BEFORE a radiation source (radioactive material) is acquired/used while registration can be performed after acquiring a machine which produces radiation (usually within 30 days following the first use)—except for accelerators and mammography units.

There are two types of licenses: specific and general. A specific license must be obtained by application to the regulatory agency. The license must be received before acquiring the radioactive material. A general license is not a document issued to the licensee but is an authorization that already exists within the regulations (state or federal, depending on which has jurisdiction) for anyone acquiring specific sources/devices incorporating radioactive material. The device must have already been approved by the NRC or an Agreement State for distribution as a generally licensed device. The licensed distributors of these devices are required to notify the NRC or Agreement State (depending on where the general licensee is located), every calendar quarter, of the distributions within their jurisdiction. Some states now require that the general licensee register the use/possession of the generally licensed device (and submit a small fee). Whereupon the state will return a "General License Acknowledgment". This information is then used by the agency to perform periodic inspections and supply information regarding safety. Generally licensed devices must be returned to the distributor when the licensee no longer desires to possess it. They cannot be directly transferred to anyone that does not possess a specific license authorizing possession of that specific device.

Preparing an application for a specific license can be quite exhaustive. It must provide general information regarding the applicant and very specific operating procedures for operations using radiation sources. The agency then issues a specific license document identifying the authorized materials and uses and listing conditions of operation. The application for a general license is actually a form of registration and simply provides information regarding the general licensee. In Texas, the agency returns a simple "general license acknowledgment" document with some operating requirements (conditions). In general, a license condition, a regulation, and a specific law all have equal weight (according to attorneys encountered by the author and a legal course taken by the author). In practice, however, there are variations and exceptions.

One should note that there is a type of specific license called a "broad license". The broad license is an authorization to perform research and development activities using fewer agency generated controls. In research and development, specific rules are not always applicable because new approaches to use of radiation may be hampered by requirements that are not quite applicable. A broad licensee must have a "radiation safety committee" to oversee the research and development activities, (usually) more stringent operating and safety procedures, a very qualified staff, and considerable material resources. Research medical facilities, research industrial companies, and large universities often choose a broad license approach. In general, the large broad licensee will issue "sub-licenses" or "sub-permits" for researchers and then control them in a manner similar to a radiation control agency. The radiation control agency, then, simply has to inspect the management of the broad license and "spot check" some of the broad licensees authorizations. A non-broad specific licensee must seek amendment of his/her license each time a new authorization is desired. A broad specific licensee has more flexibility in radiation use so the agency workload of continually amending a license is greatly reduced.

X-Ray Registrations

In general, X-ray devices must be registered within 30 days of first use, but there are exceptions. It would be prudent to check with your state regulatory agency and determine the state's requirement. There are several types of registrations for radiation sources: healing arts devices, non-healing arts devices (industrial or educational), and service providers.

Healing arts x-ray devices consist mostly of medical diagnostic x-ray and fluoroscopic equipment, dental x-ray, veterinary x-ray, chiropractic x-ray, and pediatric x-ray, although there are a few other types of uses. Mammographic x-ray equipment/ systems must be certified to meet stringent requirements.

Non-healing arts x-ray units are devices that are not used directly on humans or animals and are operated in industrial facilities, educational facilities, and even in healing arts facilities (but not related to healing arts activities).

Service providers, persons providing services to registrants, must register those services if they meet certain requirements of the Texas rules. For example persons performing service or maintenance on x-ray equipment (assemblers or installers), performing radiation surveys to show compliance, or providing "agency accepted" training courses usually must register these services with the stare.

3. Requirements of Radioactive Materials Licenses/X-Ray Registrations

Permit Application Process

Permit is a general term for license, registration, certification, etc., and may be used interchangeably in this text.

Radioactive Material Licenses

Specific License

With little exception, prior to use/possession of radioactive material, a valid specific license must be acquired. A radioactive materials license application must be submitted to the appropriate regulatory agency on an application form prescribed by that agency. Documents supporting the application are also required. The more complex or involved the use of radioactive material, the greater the supportive document requirements. The documents must attest to the qualifications of the radiation safety officer and the qualifications of the users/operators of radioactive material and related devices. Some uses have requirements specified in the radiation control regulations, while others are more generalized and are required through "administrative procedures", such as described in *regulatory guides*. The supportive documentation must also demonstrate that the applicant has established safety procedures adequate to prevent workers and the general public from being exposed to radiation that might exceed any limit, and, in many cases, will prevent "unnecessary radiation exposure". Interestingly, the regulations allow radioactive material to be owned without a specific license (under a general license). However, possession does require a specific license.

Broad License

Broad licenses are specific radioactive material licenses. The broad licensee has considerably more flexibility in operating its program, but must show a great deal of experience and knowledge in radiation safety. In a broad licensed program, a formal radiation safety committee oversees operations and approves research protocols. The broad licensee must control authorized users in the same manner as a regulatory agency controls individual specific licensees. The regulatory agency will then inspect the broad licensee's controls and some of the individual programs to determine the success of the broad licensee's own control program. Broad licenses are usually the choice for large medical and educational programs. A few large, industrially oriented, companies have been issued broad licenses for research and development.

Once an application has been received by an agency, the review process may take months or years before a license is issued. Providing either too little or too much information can lead to delays. The agency will be examining a license application and supporting documents to assure that all requirements have been met. If there is insufficient information, the agency will request additional documentation. If there is too much information, the agency reviewer will have to "weed through" the documents to assure that the applicant hasn't inadvertently "written in loopholes". In providing the optimum information to assure prompt license issuance, the "kiss" (Keep it Simple, Stupid) method usually works best. Many applicants prepare procedures that they feel the agency wants - anticipating that this will assure prompt license issuance. This slows the process down. Since an agency must issue a license if all requirements are met, it is advisable to restrict the submitted information to that which is required.

It is the author's opinion that a license application be supported by a "radiation safety manual" which includes all procedures and information relevant to the desired licensed operations. Each page should be numbered and dated. When later modification of a procedure, or addition/deletion of a procedure, is necessary, the licensee can simply submit replacement pages for the affected "old" pages - also numbered and dated to assure that the changes are obvious and visible to all and the effective dates of the changes are clear.

Generally Licensed Devices

These licenses are another exception to the application process. Such devices (containing radioactive material) are authorized and licensed through an agency's regulations. They are obtained from a licensed manufacturer and controlled through specific requirements in the regulations. The devices must meet certain criteria specified in the regulations and the manufacturer is required to provide a quarterly report identifying devices sold in an Agreement State and/or in a state under NRC jurisdiction. Many states now require registration of such devices (in Texas, the registration results in the issuance of a "General License Acknowledgment"). The NRC is in the process of implementing a similar requirement. This process allows agencies to track devices and keep them from being "lost[41] from the regulatory process".

X-Ray Registrations

Devices that produce radiation electrically (contain no radioactive material) are usually required to be registered in the state where they are acquired and used. No federal agency regulates the operation of such devices, with the exception of the uses of

[41] *Author's note: In the late 1970's, the author (an inspector for the DSHS's radiation control program at the time) was assigned to identify a number of generally licensed devices installed in Texas and inspect them. The project was intended as a survey of the condition of the generally licensed devices in the state. Fifteen (15) reported devices were selected from agency files (stretching from Dallas, to Houston, through south central Texas) and the reported information was used to identify the general licensees and the locations of their devices. Only 7 of the companies and devices were found—the other 8 had disappeared" from the business world. Of the 7, only one (1) was in substantial compliance with the requirements. The registrations covered a 15 year time period. This information was provided to the NRC and other Agreement States at several meetings by staff from DSHS.*

certain lasers and mammography devices. Registration usually consists of filing a form with the state agency. Operating and safety procedures are required by many states. In the case of accelerators, if they contain/use radioactive material - they are licensed. Otherwise, they are registered.

X-ray Machines

There are two categories of devices. One category consists of those used in the healing arts (medical, dental, veterinary, etc.) and the other category includes all others - usually referred to as "industrial devices". The basic difference is that the devices in the healing arts are designed and intended to deliver radiation to humans (or animals) while the industrial devices and are to be operated without exposing humans. This difference has caused two different regulatory schemes to evolve. Unfortunately, many agencies have lumped the two groups together in the regulatory process - making the process difficult for both the agencies and the registrants. Texas has broken out its registrations into several categories that ease the process.

Accelerators

Accelerators are normally registered, unless they contain radioactive material (target), in which case they will probably be licensed.

Radiation Safety Services

Some states, such as Texas, also require registration of radiation safety services that are not licensed, such as for radiation surveys of x-ray equipment and for radiation safety training programs. Note that, in some states, a person providing services to medical facilities may also be required to possess a "medical physicist" license B a separate qualifying process. A licensee or registrant arranging for radiation safety services should verify that the service provider meets the requirements in that state B otherwise, the obligations of the licensee or registrant may not be fulfilled.

Other Sources and Uses

Exempt Sources

One exception to the application (authorization before possession) process is for the use of exempt radioactive material. Exempt sources will be identified when the user purchases the material from the supplier. Cautions and instructions should be provided with the material. Registration of exempt sources is not required, as compared to generally licensed devices. Exempt sources can be identified in applicable radiation control rules, but the suppliers will usually have all of the necessary information. *Caution: A licensable source that decays to below exempt quantities is not automatically exempt!*

Lasers

Currently, Class IIIB, IV, and V lasers are required to be registered in Texas and other states. FDA requires that operation of high-powered lasers in shows meet certain criteria (to protect eye-sight), but operators can obtain a variance for particular circumstances—even if registered with a state program. Not all states have a registration program. If there is no state program, a laser operator should be able to determine the requirements for their state through the laser provider or the FDA.

Mammography

The federal government has implemented more stringent requirements to assure the qualifications of persons involved in providing mammographic services. The provider, physicians reading the mammograms, x-ray techs, and mammographic equipment must be certified as a "system". FDA has the federal authority for regulating this process. At the same time that the strict federal requirements were being put into place, a number of states (Texas included) were also establishing laws to provide more stringent requirements for mammographic operations. After implementation of the stronger criteria, Texas contracted with the FDA to implement a coordinated certification program. Certification is not the same as routine registration of x-ray equipment. Mammographic equipment may need to be both registered and certified - depending on the state.

Permit Information

Regulatory Guides

Regulatory guides, or "reg guides", are provided by many agencies for the more complex types of programs, such as industrial radiography, well-logging, and medical uses. Some states have them on-line. In some cases, they are provided for programs that are more numerous—but not particularly involved or complicated, such as for portable moisture/density gauges. Since it is easier to change a reg guide, agencies can modify desired procedures more readily and without going through the time-consuming rule implementation process. Reg guides also provide more flexibility in developing a safety program. The "desired" procedures are more flexible and are only binding if committed to by the applicant/licensee. Should a required procedure be impractical for a licensee, usually, an agency will accept alternatives that can be shown to provide the equivalent level of protection. If an agency insists on including a procedure that does not "fit" a applicant/licensee, an administrative hearing on the matter can be requested by the applicant/licensee. Fortunately, most reasonable alternative procedures are accepted by an agency.

License/Registration Conditions

When a radiation user's application is approved and a license or registration is issued, the work has only just begun. The successful permittee must then assure that all "promises" made during the application process will be fulfilled. The license will formally list the authorizations and the *"conditions of use".* These should be immediately reviewed to assure that authorizations include the radioactive materials and/or devices that were applied for, and that the conditions are implemented into the program. Failure to abide by a license condition is a violation. Registration certificates issued by a state may or may not incorporate conditions. They usually do identify the authorized equipment and authorized uses.

4. Requirements of Written Operating, Safety, and Emergency Procedures

Management Requirements: ALARA and the Radiation Protection Program

There are two entities that are responsible to assure a safely operating program: management and the user (worker). Management must, of course, set up the operating safety program and train the workers to work within the program requirements. The following areas should guide the radiation safety officer and the company in setting up a radiation safety program. The safety program will necessarily include ALARA and a radiation protection program (RPP).

ALARA

ALARA is defined in the Texas regulations as: *"As Low As Reasonably Achievable. This means making every reasonable effort to maintain exposures to radiation as far below the dose limits as practical consistent with the purpose for which the licensed activity is undertaken, taking into account the state of technology, the economics of improvements in relation to benefits to the public health and safety, and other societal and socioeconomic considerations, and in relation to utilization of nuclear energy and licensed materials in the public interest."*

Before the formal implementation of ALARA, licensees and registrants would only react to radiation exposure situations when a worker exceeded a regulatory limit (if then). ALARA requires that each licensee/registrant maintain constant monitoring and control of employee exposures - setting lower practical limits for the type(s) of operations and then taking action when employees exceed the company established lower limits. The licensee/registrant sets the lower limits based on exposures expected during safe operations by properly trained employees.

For example, (using TEDE's for typical gamma exposures) suppose we have Worker A and Worker B performing the same basic radiation duties and doing about the same amount of work. Worker A is receiving 20 millirem (0.2 mSv) per month (240 per year) and Worker B is receiving 400 millirem (4mSv) per month (4,800 per year). Both of these are below the regulatory limit of 5,000 millirem (50 Sv) per year. However, obviously Worker B is receiving 20 times the exposure that he/she should receive. If Worker A can perform the same work at the lower exposure rate, then Worker B's procedures and work habits need to be evaluated and Worker B needs to be trained to perform operations that maintain his/her exposures at the more reasonable level. This can be accomplished with little cost to the company. A caution for this example is that Worker A or Worker B may be not using his/her personnel monitoring properly.

Maintaining an ALARA program and monitoring its success is a radiation safety officer and management responsibility.

Radiation Protection Program

"Radiation Protection Programs" (a process) became a requirement for all licensed programs when implemented in 10 CFR 20 by the NRC in the early 1990's. Agreement States were required to implement compatible rules. The NRC's original intent was to apply the radiation protection program requirement to nuclear power plants. As implemented, however, it also applies to non-power plant radioactive materials users. Thus, for those state radiation control programs with integrated rules that apply to radioactive material users and radiation machine users alike, the RPP requirement became applicable to many radiation machine users, as well. This is the same mechanism that provides standard dose limits, and other radiation protection standards, for radiation machine users. While standardization in many areas of radiation protection for all radiation users in all states can be beneficial to users and regulators, alike, it is not necessarily beneficial or practical in some areas.

In 1996, the Texas radiation control program sent a letter to the NRC requesting clarification. The basic question presented was "was not the RPP requirement intended only for nuclear power plants and related licenses and not for materials licenses; and, should it be a compatibility requirement for the Agreement States"? After a period of prodding the NRC, the Texas program received an NRC letter indicating that the requirement was applicable to all licensees. The Texas program had already implemented the requirement in rule, as shown in Table 5-2.

Radiation Protection Program Requirement
25 TAC §289.202: (e) Radiation protection programs. (1) Each licensee or registrant shall develop, document, and implement a radiation protection program sufficient to ensure compliance with the provisions of this section. . . . (2) The licensee or registrant shall use, to the extent practicable, procedures and engineering controls based upon sound radiation protection principles to achieve occupational doses and public doses that are as low as is reasonably achievable (ALARA). (3) The licensee or registrant shall, at intervals not to exceed 12 months, ensure the radiation protection program content and implementation is reviewed.

Table 5-2: Radiation Protection Program Requirements

Operating, Safety, and Emergency Procedures (OSEP)

From the early stages of radiation control, safety procedures have been a standard requirement. Each license applicant was required to present an acceptable "operating, safety, and emergency procedures" (OSEP) with the application. To some extent, this requirement spilled over into the radiation machine registration process. The RPP elements are usually already addressed in a licensee's OSEP, so the RPP is generally redundant (it would have been simpler to assure that the RPP elements are addressed in each licensee's OSEP). The RPP is basically a subset of procedures in the typical OSEP. Many regulators labor under the mistaken concept that an RPP is a document. Actually, an RPP is a program. The easy course of action in implementing an RPP, then, is to create a simple document describing implementation of such a program and listing the sections of the existing OSEP that address the elements of an RPP.

The minimum elements of a radiation protection program, if one addresses the overall requirements of TAC §289.202, should address:

- Personnel monitoring requirements and occupational dose limits
- Radiation surveys and instrumentation
- Access controls for radiation areas
- Respiratory protection
- Security of radiation sources (storage/use)
- Posting of areas and rooms
- Public dose limits
- Labeling of containers
- Receipt of packages containing radioactive material
- Waste storage, processing, transfer and/or disposal procedures
- Management of required records
- Reports of incidents

Table 5-3 Radiation Protection Program Elements

All of these should be addressed in an OSEP for each license and should be covered in all new license applications. If a given area is not applicable to a program, the OSEP should identify the area and indicate that it is not incorporated - providing justification for not incorporating a procedure. Some more complicated or involved radiation use programs may need to address other areas of the applicable rules.

General Safety Procedures

In preparing an OSEP, the following areas are generally the responsibility of the radiation safety officer and company management. If a procedure is not applicable to the program, it should be clearly indicated (an agency may disagree). A workers' operating and safety procedures document or manual should be developed based on the general procedures supporting the licensed or registered operations.

Management Procedures

Management must address the following procedures, as applicable. (generic procedures using these elements are listed in Appendix 4)

- General Safety Policies
    < Management structure and controls
    < RSO Authority
    < ALARA policy
    < Radiation protection program and annual review
- Radiation Monitoring
    < Personnel monitoring and bioassays

  - Restricted/unrestricted area monitoring
  - Bioassays
- Posting of Areas and Rooms
- Labeling of Containers
- Posting of Notices to Workers
- Radioactive Material Receipt Procedures
- Storage of Radioactive Material
  - Storage security
  - Storage procedures
- Radiation Surveys
  - Storage facility radiation surveys
  - Use area radiation surveys
  - Vehicle radiation surveys
- Radiation Survey Instruments
- Radioactive Material Inventory (physical)
- Records Management
  - Use/Storage log
  - Receipt and transfer records
  - Leak test records
  - Personnel monitoring reports
  - Instrument calibration records
  - Radiation survey records
  - Training records
  - Inspection and maintenance records
  - Inventory records
  - Transport records
  - Use radiation survey records
- Transport of Radioactive Material
  - Transport modes
  - Vehicle placarding
  - Shipping papers/manifests/bills of lading
  - Vehicle radiation surveys
  - Transport procedures (proper packaging and labeling)
- Equipment inspection and maintenance procedures
  - Equipment inspection
  - Routine Maintenance
- Emergency Procedures
  - General field procedure
  - Theft of a source/device
  - Loss of a source/device

< Damage to a source/device
< Exposure of non-radiation workers
< Emergency during transport
< RSO Responsibility (investigation, correction, notification and reports to agency)

- Leak Test Procedures
- Controls and Procedures for "Loose" Materials
- Contamination Controls
- Decontamination Procedures
- Waste Disposal (methods, resources, precautions)
- Employee Training
  - < Inexperienced new employees
  - < Experienced new employees
  - < Annual review of radiation workers
  - < Prohibitions from conducting unauthorized functions
- Transfer Authorization Procedures
  - < Transfer of equipment to another party
  - < Acceptable methods and/or documents showing authorization

Worker Procedures

A radiation worker safety procedures manual should be developed (and made available to the workers) and should include the following. (generic procedures using these elements are listed in Appendix 4)

- Employee/Radiation Worker Responsibilities and Duties
  - < ALARA
  - < Prohibitions from conducting unauthorized functions
  - < Authorizations required before conducting work
  - < Employee responsibilities
- Personnel Safety
  - < Personnel monitoring requirements
  - < Personal safety (required protective equipment)
- Storage Security
  - < Employee responsibility
  - < Storage procedure
- Transport Procedures
  - < Loading vehicles for transport of radioactive material
  - < Procedures used during transport
  - < Shipping record
- Safe Operation Procedures

< Operations using radiation sources
< Operations at field sites
- Records
- Emergency Procedures
  < Securing the area
  < Contacting the RSO
  < Damage inspection
  < Actions for minor damage
  < Actions for serious damage
  < Records

An OSEP that includes or addresses the above listed procedures (both lists) should adequately support a license application and should be acceptable to the regulatory agency with a minimum of "fine tuning". Please note that the list is not all inclusive. The regulatory agency will probably contact you for more information or clarification at least one time during the licensing process.

5. Training of Workers

Purpose of Training

Studies have shown time and again that proper safety training for workers reduces accidents and injuries and that these reductions benefit industry by reducing costs. Most accidents can be prevented by making workers aware of the hazards associated with their duties and providing them with knowledge and procedures that will allow them to protect themselves. For new, untrained workers, two (2) types of training are usually required for radiation use: (1) basic radiation knowledge and safety training and (2) on-the-job-training (OJT) which allows an inexperienced worker to benefit from an experienced worker's knowledge and skills. Even experienced workers that are newly hired or reassigned to radiation related duties will usually need refresher radiation safety training and OJT in the specific procedures that they will be using. Thus, radiation control programs have become more sensitive to training issues and now demand more formal training protocols.

Regulatory Requirements

In general, any training program will be required to include classroom-type instruction in basic radiation knowledge, radiation hazards, and self-protective methods. In addition, they will need to include development of hands-on experience through

OJT. Licensees and registrants can provide both of these through "in-house" training programs if they can show that the classroom training will be conducted by qualified instructors and that OJT will be conducted under the supervision of qualified Individuals. They can also have a program of providing for classroom training through an agency approved/accepted training program and then provide the OJT in-house.

Agency Accepted, Approved, Authorized Courses

The NRC and Agreement States require that classroom training be provided through a course "approved by the NRC or an Agreement State". Training criteria for industrial radiography, well-logging, and some medical uses is specifically listed in regulations (training hours, training content, exams, etc.), so the regulatory agency has clear authority for evaluating and approving training programs for these activities. Training for most other areas of radiation use is not very specific, so training content and methods for other uses allows for more flexibility. An approved or accepted training course will be authorized through either a license or a registration. Some companies and consultants have applied to the Texas program to be listed on the agency's website along with the above accepted courses.

Training Programs

General subject areas required for training programs (in the regulations of the NRC and some Agreement States) can be found for industrial radiography and well-logging. Some states' regulations also address naturally occurring radioactive material (NORM) training areas, as well. If not found in regulations, training criteria may be found in regulatory guides issued by the agency with jurisdiction. A licensee should develop a company training program and include it with the license application as it will probably be required. This is not always cost effective for small companies. It is generally better to send workers to a service company course for general radiation safety training and set up a basic on-the-job training regimen for internal, practical training. In Texas, "agency accepted" training courses are only those for industrial radiography and well-logging because those are the only two types of use that have specific training criteria listed in the rules. Training courses for other types of use, such as portable moisture-density gauges, spinning pipe gauges, RSO, etc., are placed on a list as "appropriate" training courses that are accepted for purposes of meeting training requirements. Since not all training entities are evaluated prior to being listed, a license applicant or licensee should be particularly careful in selecting training resources. Acceptance by the regulatory agency should be assured before spending time and money on training of personnel - only to find that the training is not acceptable.

Training Resources

When planning a training course, standard training resources and techniques can be used. Most courses today rely on video tapes, overhead projections, slides, and "show and tell" objects. An instructor must be qualified by training and experience.

Unfortunately, "state of the art" training videos are difficult to locate and acquire. Several are made by service companies, but they are generally specialized. There are some old movies made by the military/federal government that have been transferred to video tape but the video quality is often poor. Diligent research may turn up videos that can supplement a training program. One can also record videos of operations and equipment - which may be particularly useful when the radiation related equipment is not available at the training site.

Overhead projections can easily be prepared by the licensee or instructor. Computer overhead projections are beginning to be used more and more and can make a training program more interesting. The same can be said of slide projections. Slides can also be added to a computer overhead projection scheme.

"Show and tell" objects can consist of a quite a number of items. When discussing radiation survey instruments, demonstrating the instrument's operation with a check or calibration source sends a better message than simple discussion or pictures. Mock-ups can be made of devices that hold radioactive material to demonstrate their inner workings without exposing the trainees to radiation. Non-radiation producing items, such as safety devices (for example, collimators) and personnel monitoring devices can be used to familiarize the trainee with tools that will be used in radiation work. The instructor should provide the trainees with devices that will help them gain a good understanding of the radiation protection principles being presented.

Training Course Content

The areas to be included in training courses can be obtained from the applicable rules of the agency, from regulatory guides, or from general information that the agency may provide.

A general list of topics will probably include (using Texas rules as a guide):

1. Radiation Fundamentals
    - characteristics of radiation
    - units of radiation dose (rem) and activity of radioactivity (curie)
    - significance of radiation dose

  - < radiation protection standards
  - < biological effects of radiation
- levels of radiation from sources of radiation
- methods (principles) of controlling radiation dose
  - < time
  - < distance
  - < shielding
- radiation safety practices, including prevention of contamination and methods of decontamination
- discussion of internal exposure pathways

2. Radiation Detection Instrumentation
   - radiation survey instruments
     - < operation
     - < calibration
     - < limitations
   - survey techniques
   - individual monitoring devices

3. Equipment to be Used
   - handling equipment and remote handling tools
   - sources of radiation
   - storage, control, disposal, and transport of equipment and sources of radiation;
   - operation and control of equipment
   - maintenance of equipment

4. Requirements of Pertinent Federal and State Regulations

5. Licensee's Written Operating, Safety, and Emergency Procedures

6. Licensee's Record Keeping Procedures

This is not an all inclusive list, however, the particular requirements and recommendations of the agency should be addressed. Appendix 2 provides lists for certain types of operation according to Texas rules and reg guides.

6. Records Maintenance Procedures

Functions of Records

Every licensee questions the necessity of a given record at one time or another. Behind each type of record, there is probably a story of an accident or severe non-compliance. Why keep all of these records? Certain records of a radiation safety program provide two primary functions:

(1) they show that the licensee/registrant has performed a safety function and documented the results; and

(2) they allow the agency to inspect the records and find some assurance that a required safety function was performed by the licensee/registrant.

A third function (and when it happens it is an important function) is that the records often provide essential information when an incident occurs - such as a lost source. In this case, properly maintained inventory and use logs can indicate when the source was last removed from storage, who removed/used it, where and when it was used, and when (or if) it was returned to storage. Agencies use these records when investigating cases of lost or missing sources and they have been essential in recovering a number of sources. Licensees/registrants can use them in the same fashion.

Types of Records

Each required safety function and each action supporting safety or the license/registration will probably have a record that will document the function or elements of the function. A record storage location should be indicated in the license or registration application. The records should be accumulated there and controlled by the radiation safety officer. Common records are (there can be many more):

Use/Storage Log (or Utilization Log) — a record which documents the removal of a radiation source from the storage facility and indicates the device's model and serial number; the person responsible; the date of removal; the location where the source is used; the date/time the device was returned to storage; and the identification of the person returning the device to storage (a different person may return it). A radiation survey of the storage facility may also be documented.

Receipt and Transfer Record	documents the receipt, transfer, and/or disposal of radiation sources. The record should identify (must "uniquely identify the source of radiation") the source's model and serial number, quantity, include the date and the identification of individual(s) making the record. May be combined or separate records.
Leak Test Record	the service company's (or licensee's, if authorized) certificate which uniquely identifies the source and indicates the date of the test and the results. The leak test interval is determined by the date of the swipe - since the various service companies take differing times to process the samples and return the results.
Personnel Monitoring Report	the service company's provided record for film badges, TLD's, etc. May also include daily pocket dosimeter readings. Should also include the history form and reports to former employees of exposures.
Instrument Calibration Records	the service company's (or licensee's, if authorized) radiation survey instrument calibration certificate. May also include pocket dosimeter calibration, on-site calibration of multi-channel analyzers, etc.
Radiation Survey Records	includes, as applicable to the program, records for radiation surveys of the storage facility at each authorized site, use areas, vehicles (if the vehicle is used for storage) and devices. May also include surveys of areas where contamination may be present (wipe surveys).
Public Dose Record	Records demonstrating compliance with public dose limits pursuant to surveys and/or calculations.
Training Records	records showing the qualification and training of each authorized user. They may include "trainee" records for certain types of uses, such as industrial radiography. These records will vary from company to company—being established during the licensing or renewal process.

Device Inventory, Inspection, Maintenance Record	a record used to document the presence and physical inspection of each radiation source or device, and the performance of any necessary maintenance. In some cases, may be combined with inventory records. Not all types of use are required to perform and document this function.
Radiation Protection Program Annual Review Record	a record of the annual review of the safety program. This record will depend on the content of the radiation protection program.
Radiation Worker Annual Evaluation Record	provides for a record and annual review of radiation users for continued assessment of each worker's proficiency in radiation use and safety. Not required for many uses—but strongly recommended for most.
Training and Audit Records	a collection of all required training records and all individual audit records. Very company dependent.

If work is performed at temporary job sites, the following records, or copies of documents, usually must be maintained and available and at each temporary job site:

Authorization Document	a copy of the radioactive material license and/or x-ray registration certificate.
Personnel Qualification or Certification Record	a copy of any required certification of qualification, as in a radiographer's ID card.
Radiation Safety Manual or Operating and Safety Procedures	a copy of the company's procedures that direct the radiation workers on how to safely perform the work and recover from emergencies (accidents, incidents).
Regulations	copies of the code, regulations, rules, etc., that are applicable to the type of work with radiation. These are usually determined by a license condition or the company's procedures.
Radiation Survey Records	radiation survey records for the period of operation at the job site should be available. Could be use area, vehicle, etc.

Personnel Monitoring Records	pocket dosimeter records for the period of operation at the job site—if required. For some types of use, such as industrial radiography, recording the starting and ending readings of direct read-out monitors is required.
Radiation Safety Equipment	the current instrument calibration and leak test records for devices at the job site are usually required.

Great care and attention should be given to developing records that will be required for documenting the program. Avoid duplication of functions across records. Combine functions to reduce the number and types of records as much as possible. During regulatory inspections in the past, the author has seen licensees have as many as four records documenting one function - such as inventory or device inspection. Once a record is submitted to support a license or registration - it becomes part of the license and omissions become violations if noted during inspections. A little extra time spent here can save a great deal of distress later.

Records Maintenance

For many years a "paper" requirement, many records can now be maintained on various media. Computer magnetic storage, fax printouts, and micro fiche, are few of the more recent media authorized. If a media other than paper is planned, the agency's rules should be reviewed to assure that the alternative is allowable. Also, the records must be made available during inspections so the alternate system should be set up so that an inspector can review them without requiring a lot of time and effort—unless you want an irritated inspector present in your offices for much longer than necessary. Most companies use paper with a "standard" folder filing system. (Note: design records to fit a standard 8.5" X 11" sheet of paper so that they will be easily faxed and/or copied.)

Filing and Retention of Records

During inspections, the inspector will usually review records in an order that groups similar records. For example, personnel exposure histories, current personnel monitoring records, and pocket dosimeter summaries might be reviewed together. If they are filed together, the inspection should proceed more quickly. They will also be easier to review during the company's quality assurance process. Time periods for retention of most types of radiation safety related records are now specified in the rules of the NRC and Agreement States. For records not specified by rule, be sure to indicate the retention time in the radiation safety manual or operating and safety procedures. Otherwise, they may be required to be maintained "forever".

Government Record

Records, as discussed in Chapter 4, documenting radiation source information and conductance of safety procedures have a number of important functions in the safety process. However, it should be noted that a government record is often defined as "a record maintained by a person to fulfill government requirements". It is not necessarily a record of a government agency. Further, the laws of many states prohibit "tampering" with government records. Thus, alteration or fabrication of records required by an agency could lead to legal problems that are not necessarily directly related to radiation safety and compliance.

7. Agency Inspections

Purpose of Agency Inspections

When a government agency issues an authorization for a person or company to conduct some business or operation, it does so because the agency has been authorized and required to do so by law. The enabling legislation almost always requires that agency to control the authorized activities and assure that the activities comply with all requirements. The most effective method to assure compliance is to send trained representatives to audit or inspect the program. The following tools are generally used by radiation control programs:

- Audit—a detailed review of all required and associated radiation related records, careful inspection of all radiation related equipment—including inspection and review of shielding of use and storage facilities, observation of workers procedures (while working), inspection of all field operations and sites, etc.
- Inspection—a detailed review of representative records, review of some workers procedures (while working), inspection of field operations and sites.
- Survey—a review of selected radiation safety records and radiation surveys of equipment - particularly x-ray equipment.
- Self-inspection—an agency form, to be filled out by the permittee, showing the primary safety requirements are being met.

- Field inspection—inspection of field operations - usually at temporary job sites.
- Investigation—a detailed review of circumstances relating to radiation incident or accident; or a complaint of noncompliance; or a worker's complaint. It is usually narrow in scope and involves audits of relevant records, inspection of equipment, and interviews of personnel. A full inspection often is performed at the time.

Inspection Schedules

Inspection schedules for different types of use of radiation sources are generally set by the relative degree of radiation hazard and complexity of the radiation safety requirements. For example, industrial radiography has very detailed requirements for operations and training of personnel. It also has a history of serious accidents. Schedules should be set for no less than an annual inspection for this type of operation. Gas chromatography, on the other hand uses very innocuous sources. Exposure risk is extremely low and there are few reported incidents related to their use. Inspection schedules may be set for 4 or 5 years.

The NRC sets overall inspection schedules for its own activities and requires that Agreement States set schedules that are at least as frequent for their own issued licenses. The NRC periodically reviews the Agreement State programs and one of the areas of review includes the states' inspection schedules and whether the states are keeping up with them. A licensee/registrant can request an inspection schedule from an agency under the states (or federal) open records provisions.

The following intervals are examples of those that may be set for the indicated type of use <u>for licenses</u> (recommended by the author):

- 6 months—radioactive waste storage and processing, major producers/processors of unsealed sources, and active large-scale uranium mines.
- 1 year—distribution, industrial radiography, well logging, broad licenses, sterilizers, nuclear pharmacies, large medical, and teletherapy.
- 2 years—portable gauges, medical diagnostic and therapy.
- 3 years—veterinary uses, sales, and consultants.
- 4 years—fixed gauges, gas chromatography, generally licensed devices.

The above is not an all-inclusive list.

Inspection Procedure

As previously mentioned, **records** related to radiation safety and the license or registration are inspected. Depending on the size of the program, the inspector may review all records or may review only a representative sample of the them. The records will tell a story. A licensee that has them all together and available may find that the inspector will skim through them quickly. On the other hand, a licensee having difficulty promptly finding and producing the records may be in for a long inspection. Poor record management might be an indication to the inspector that the safety program is not being properly maintained and the inspector may ask to review every record.

Records generally inspected are:

< **Radiation sources**—use/storage logs, radiation source receipt and transfer records, transport records, inventory, inspection, and maintenance, and radioactive material leak test records.

< **Radiation safety/protection**—personnel monitoring reports, instrument calibration records and certificates, radiation survey (use, storage, and vehicle) records, worker training records, and annual radiation protection program review.

< **Authorized sites**—other locations authorized by the license/registration and records of operations conducted at field sites (temporary job sites).

< **Miscellaneous records**—well logging agreement (agreement among principals associated with a well on actions to be taken if source is stuck down-hole or well becomes contaminated), property owner agreement/acknowledgment (indicates that property owner knows that radioactive material will be stored or used on his/her property), reciprocity notifications to NRC or other states.

< **Other**—copies of appropriate parts of applicable regulations, radiation safety manual or emergency operating and safety procedures, and copies of previous agency notices and records of corrective actions.

Facilities—During inspections, the following physical items will probably be reviewed:

< **Use areas**—the inspector will visit areas where radiation sources are used, examining posting and labeling, checking controls and interlocks, and performing radiation surveys in surrounding unrestricted areas. Worker stations will also be reviewed to assure that no one is being unnecessarily exposed.

< **Storage areas**—will be examined in the same detail as the use areas. If area monitoring is used or required, it will be reviewed for appropriateness.

< **Transport vehicles**—will also be inspected for proper security of radioactive materials during transport, appropriate labeling of transport packages, and placarding of the vehicle.

< **Use at temporary job sites**—will be reviewed to examine how overnight storage is accomplished.

Personnel Training—An inspector will not only review the training records of workers, but will also interview the workers to verify that they did receive the training indicated on the records. In addition, a field exam is often given for industrial radiographers.

Miscellaneous—in the event of a radiation incident, during the course of the investigation a full inspection is often performed - in many cases at the audit level. Persons outside of the licensee/registrant's control may be contacted and interviewed. This is usually outside of the routine inspection interval.

Management Summation—prior to leaving the facility the inspector will notify company management (above the RSO) of the inspection findings and the type of written notice to expect from the agency. Some agencies provide a brief form indicating that no violations were found or providing a check list of some common, but low severity level, violations. In this case, no further correspondence between the two parties occurs except that the agency may request written correction of the "checked" violations.

Inspection Findings

Results of Agency Inspections

Fortunately, the inspector does not have the final word on the inspection. The inspection findings (inspector's report) should be reviewed by other experienced agency personnel to assure that inspection practice and policy have been followed. Any information reported back to the inspected party would then be more accurate. Further notices to the inspected party should be supplied by the radiation control office.

Agency Notices—Most agencies will attempt to provide the inspected party with a written notice of the inspection findings (generally called a "notice of violation") within 30 days. This is not always possible due to the size of the radiation control program or geographic size of the state. The inspector will usually indicate the time frame. If a licensee/registrant doesn't receive a notice within a reasonable time, it might be prudent to call the agency and inquire. Occasionally, an inspector might indicate a violation was found that, upon further agency review in the office, is found not to be a violation. The written notice will be the final indicator of what needs to be corrected or addressed.

< **Correction of Violations**—Serious violations should be corrected immediately - not when the written notice is received. The notice will list the violations, indicate that they need to be corrected, and specify that the corrective actions need to be reported to the agency by a certain date (usually 30 days after the date of the notice). It will probably ask that the date of corrections be indicated.

< **Responding to Agency Notices**—Care should be taken to respond to the notice precisely as requested in the notice. If corrections cannot be accomplished and a written response provided to the agency within the specified time period, simply call the agency and request an extension of time. The agency would prefer the corrective process be conducted properly and completely than rushed and haphazardly.

Agency Enforcement Actions

A question often asked is: "what will happen if I commit a violation of the law, regulations, or license requirements"? The answer is you will either face a "mountain of paperwork" and/or a monetary penalty, and/or incarceration, and possibly considerable embarrassment. This can apply not only to the licensee, but also to the employee/operator. Generally, the licensee is responsible for all violations—but agencies are finding and developing the authority to additionally penalize employees who commit violations willfully or neglectfully.

Enforcement Processes

Possible enforcement actions that may be encountered include (but are not limited to):

- **Agency Order of Impoundment**—the agency with jurisdiction impounds a radiation source to prohibit its further use. It may be impounded in place (locked by agency lock within the licensee/registrant's facilities), or it may be physically removed and stored under lock and key by the agency. This is an emergency order and so the agency must specify that an emergency exists when invoking this type of order.

- **Agency Order of Cease and Desist**—the agency with jurisdiction prohibits the use of radiation sources until certain conditions are satisfied (such as correcting a serious radiation hazard). This is also an emergency order and the agency must specify that an emergency exists. It may also be combined with an impoundment order.

- **Agreed Order**—the agency and a party shown to be in violation jointly sign an order which lists the steps to be taken by both parties to correct the problem and satisfy all requirements. (Remember, an agency is responsible to the public and would be remiss in its duty if it did not "take to task" a party that posed a serious threat to the public health and safety.)

- **Enforcement Conference**—whereby an agency schedules a conference to meet with the licensee/registrant's representatives to seek correction of continued and/or serious compliance problems. Often, an agreement order is the tool used to complete the action, but it may lead to the more stringent actions listed in the next section (monetary penalty). This may also be termed a "management conference".

Most of the above activities will be taken only as a "last resort"—when other (perhaps gentler) options have been exhausted. However, negligent loss of radioactive material, deliberate exposure of an individual to radiation, or willful violation will most assuredly lead to an enforcement action.

Penalties

When an agency exhausts all opportunities for achieving compliance through routine inspection and enforcement activities, it may have to turn to a real "attention getter"—monetary fines and penalties!

Administrative Penalties—monetary fines can be charged by an agency "administratively". In other words, the agency can propose a penalty to be paid by the violator without involving the appropriate court system. In Texas, a party with one or more "Severity Level I" or "Severity Level II" violations can be considered for initiation of proceedings for such a penalty, but Severity Level III, IV, and V violations can cause additional penalties to be proposed and added on. The Texas agency prepares a "Preliminary Report for Administrative Penalty" and then offers an administrative hearing. A DSHS Deputy Commissioner is the final decision maker for those that request a hearing on the matter. Often, the agency and the violator reach an agreement on the amount of the penalty(s) and sign an agreement order to resolve the issue. The basic penalty amount is:

> "... an amount not to exceed Ten Thousand Dollars ($10,000) a day for a person who violates the Act or a rule, order, license, registration or regulation issued under the Act. Each day a violation continues may be considered a separate violation for purposes of penalty assessment."

The levels of violation are charged by taking the percentages, shown in Table 5-4, of the above amount. All of these are subject to change at any time and are applied strictly according to the agency's current procedures. Since these are Texas rules, differences can be expected in other states.

Base Amounts	
Licensees, registrants, or certified industrial radiographers	$5,000
Persons not licensed, registered, or certified	$10,000
Severity Level I	100%
Severity Level II	80%
Severity Level III	50%
Severity Level IV	15%
Severity Level V	5%

Table 5-4 Percent of Penalty by Severity Level

Civil Penalties—most, if not all, radiation control agencies can also use civil penalties exacted through the state's court system as part of the penalty process. Usually, a violator must have already committed the same offense before a civil penalty is sought. This approach usually involves the state's attorney general and/or a county or district attorney. This may also be at the "criminal offense" level where the monetary penalties can be higher and incarceration can be part of the penalty. There have been at least two such cases in Texas.

CHAPTER SUMMARY

While the process of developing a radiation safety program, implementing it, performing all required safety functions, and getting satisfactorily through agency inspections can be a tedious one, the successful licensee/registrant is the one that properly prepares him/herself and workers, and then maintains a careful schedule of attending to all requirements. The costs associated with not achieving compliance can be great and the "recovery" time considerable.

Chapter 6

Radiation Safety Officer Duties and Permit Procedures

1. Introduction

The radiation safety officer is the key person in any program that uses radiation. In general, radiation control regulations require the identification of a qualified person who will assure that all radiation safety functions are carried out, no matter how trivial they may be. This chapter addresses the procedures that the radiation safety officer will need to attend to.

2. The Radiation Safety Officer

Radiation Safety Officer
Qualifications, Duties
The Radiation Safety Officer, often referred to as the "RSO", is the person assigned by the licensee or registrant to assure that all regulatory requirements and all radiation safety requirements are met. The RSO usually is the person that deals directly with the regulatory agency on all matters relating to the license or registration. The RSO's supervisor is also an important position in the system as the regulatory agency will go to that person if the RSO fails in fulfilling his/her duties.

The rules require that an RSO be designated for every license issued by the agency. A single individual may be designated as RSO for more than one license if authorized by the agency. Large educational and medical facilities, which may have several licenses, often assign one person to be the RSO for all permits, licenses and registrations. Facilities with broad licenses will be required to establish a "radiation safety committee" (RSC) to oversee the radiation safety obligations. The RSC will govern the operations usually carried out by the RSO.

Radiation Safety Officer Qualifications[42*]

The overall qualifications required of RSO's can vary according to the regulatory body with jurisdiction. The basic or minimum qualifications are:

- possession of a high school diploma or a certificate of high school equivalency based on the GED test
- completion of the training and testing requirements specified in this chapter for the activities for which the license application is submitted
- training and experience necessary to supervise the radiation safety aspects of the licensed activity.

There will be additional requirements for certain licenses, such as: industrial radiography, well-logging, broad (scope), and most medical licenses. These will be addressed during discussion of the licensing process.

Radiation Safety Officer Duties

The specific duties of the RSO include, but are not limited to, the following:

Operating, Safety, Emergency, And ALARA Procedures

Establish and oversee operating, safety, emergency, and as low as reasonably achievable (ALARA) procedures (OSEP), and to review them at least annually to ensure that the procedures are current and conform with regulatory requirements. This is an important and primary duty of the RSO. Properly done, the RSO will probably not have very many regulatory problems nor embarrassing incidents. A well organized OSEP that conforms to the rules is an important step in the licensing process. If too few requirements are addressed, license issuance following application can take a very long time. If unnecessary procedures are adopted, compliance will be a major effort. Too many times an applicant will include unnecessary procedures in their application or OSEP because they mistakenly think that the agency will be more amenable to issuing the license promptly. All procedures should be appropriate and straightforward.

42 * *The specific requirements will be prescribed by the state or federal agency with jurisdiction.*

Training

Oversee and approve all phases of the training program for operations and/or personnel so that appropriate and effective radiation protection practices are taught. A second, very important element of a solid radiation safety program is training for workers and all persons involved in the use of radiation. Most accidents, after investigation, that are not due to equipment failure are found to be due to inadequate training of the radiation worker. In some use areas, such as industrial radiography, well-logging, and medical, for example, specific training elements are required and courses must be approved by the agency. Periodic refresher training should also be incorporated, although not always required by the rules.

Radiation Surveys and Tests

Ensure that required radiation surveys and leak tests are performed and documented in accordance with the rules, including any corrective measures when levels of radiation exceed established limits. The appropriate types of radiation surveys and records are determined during the licensing process. Radiation surveys usually consist of radiation level surveys of use areas, storage areas, and transport. There are also surface contamination surveys that may be needed in the same areas. Leak test procedures were discussed in Chapter 3. How does one stay on the required schedules? Some RSO's simply mark their paper calendar. Others, perhaps the more successful, make use of the calendar software of the computers that sit on nearly every desk. The rest seem to periodically be corresponding with the agency regarding notices of violation.

Personnel Monitoring

Ensure that individual monitoring devices are used properly by occupationally-exposed personnel, that records are kept of the monitoring results, and that timely notifications are made according to reporting requirements for overexposures. Monitoring systems are important for both assessing personnel exposures and tracking personnel work habits. An employee consistently receiving higher exposures than other workers doing the same work needs to have his/her work procedures reviewed. Higher exposures often show the worker's indifference to the requirements and to his/her own well-being and such an employee may be an "accident waiting to happen". Identification of such employees and retraining may prevent major compliance problems. The time limits of reporting should be adhered to.

Investigate Radiation Incidents

Investigate and cause a report to be submitted to the agency for each known or suspected case of radiation exposure to an individual or radiation level detected in excess of limits established by the rules and each theft or loss of source(s) of radiation, to determine the cause(s), and to take steps to prevent a recurrence. Prompt investigation and correction of any causative problems helps to mitigate the actions that an agency may take for violations that are involved in radiation incidents. The agency will probably look closely at proposed methods for preventing recurrence. Reporting requirements are discussed in Chapter 4, Section 4.

Releases to the Environment

Investigate and cause a report to be submitted to the agency for each known or suspected case of release of radioactive material to the environment in excess of limits established by the rules. Releases that remain onsite are fairly easy to handle. Releases offsite create additional problems since communities become quite concerned when hazardous materials blow into their areas. Again, the time limits of reporting should be adhered to.

Management and Administration Policies and Procedures

Have a thorough knowledge of management policies and administrative procedures of the company holding the license. As the interface between the licensee and the agency, the RSO must have a complete knowledge of the company's management policies and must be able to effectively deal with the administration.

Assume Control in Emergency Situations

Assume control and have the authority to institute corrective actions, including shutdown of operations, when necessary in emergency situations or if unsafe conditions exist. Again, the RSO is the key person that, in some circumstances, needs to act quickly with authority to minimize any danger that may have arisen. This authority should be clearly spelled out in the OSEP, or by documents submitted with the license application.

Records Maintenance

Ensure that records are maintained as required by the rules. An astute RSO should have little trouble in this area if he/she sets up a good filing system and uses a computer calendar system to give reminders of procedures that are due. Training and using an experienced worker can also be very helpful.

Transport Procedures

Ensure the proper storing, labeling, transport, use and disposal of sources of radiation, storage, and/or transport containers. Not only must transport requirements be met during day-to-day operations, but if sources are transported to/from the supplier, then transport requirements must also be met for these shipments. Occasionally, the US DOT will send inspectors to NRC or AS licensee facilities to inspect transport records.

Inventories

Ensure that inventories are performed in accordance with the activities for which the license application is submitted. The inventory should assure that only the authorized radioactive materials are possessed and that they are the correct quantities and device models. Inventory time periods and records content are usually set during the licensing process.

Physical Inventories

Perform a physical inventory of the radioactive sealed sources authorized for use on the license every six months and make and maintain records of the inventory of the radioactive sealed sources authorized for use on the license every six months. Some inventory time periods are set by rule. The regulatory agencies expect that the radioactive material and/or its device or container will be physically inspected for its presence and the observations documented. "Paper" inventories, whereby assumption of presence is made, will lead to possible loss of control of the material. Records should include the following information, as a minimum:

- isotopes
- quantities
- activities
- date inventory is performed
- location
- unique identifying number or serial number
- signature of person performing the inventory/making the record

Personnel Compliance

Ensure that personnel are complying with the rules, the conditions of the license, and the operating, safety, and emergency procedures of the licensee. This can only be accomplished by proper training and periodic retraining of radiation workers and periodic review of their work procedures as they work. This is not a place to take shortcuts.

Agency Contact Person

Serve as the primary contact with the agency. This is a very important function. Prompt communication with the regulatory agency's staff will often mitigate a building problem, whether it be a radiation incident or a compliance problem.

Knowledge of Security Requirements

Have knowledge of and ensure compliance with federal and state security measures for radioactive material. Since 9/11, radioactive material has become a growing area of concern for all. Chapter 7 is devoted to security issues.

In order to master the above, the person assigned to be RSO for their company should seek a training course for RSO's and then continually review applicable rules and the OSEP. While regulatory agencies may not be able to recommend any particular training course, they may be able to identify those in business in their jurisdiction. In general, the material for training a radiation safety officer (basically, the material covered in this book) for most types of licenses can be covered in two (2) days. Industrial radiography, well-logging, and certain medical uses will probably require additional training. These duties should be clearly spelled out in the license application and OSEP as the regulatory agency will be examining the application to assure that they are addressed.

Assistant Radiation Safety Officer

It is recommended that an Assistant Radiation Safety Officer (ARSO) be incorporated into the safety program. A qualified, experienced person will be helpful in many areas. When authorized, an ARSO can be assigned specific duties to assist the RSO and to assure that requirements are met. For example, the ARSO can perform inventories, leak tests, and other periodic functions in the absence of the RSO. Many violations occur when the RSO is working out of town, on vacation, etc. The functions that the ARSO will be permitted or required to perform should be clearly listed in the license application and/or OSEP and the person should be fully trained to perform the duties.

3. Radiation Safety Committee

Broad (Scope) License

Many academic institutions and large companies with multiple, extensive uses of radioactive material choose to apply for a broad scope authorization for research and development. This type of license allows the licensee to set up its own "licensing process" for users in the organization. In order to meet the requirements for issuance of such a license, the applicant must require more credentials of the RSO and must also establish a Radiation Safety Committee. Some large specifically licensed operations choose to establish an RSC even though they do not possess a "broad license". This usually has been seen to be beneficial. The RSC qualifications and duties would then be as designated in the submitted license application and procedures. The duties and obligations would not necessarily need to be confined to the broad license requirements.

Additional RSO Requirements

An RSO under a broad license must have not only the credentials previously discussed above, but must also have (Texas rules):

- A bachelor's degree in an appropriate science plus 4 years of applied health physics experience in a program with similar radiation safety issues;
- A master's degree in an appropriate science plus 3 years of applied health physics experience in a program with similar radiation safety issues;
- A doctorate degree in an health physics or radiological health, or certification by one of the health physics/radiological health boards, plus 2 years of applied health physics experience in a program with similar radiation safety issues; and
- Equivalent qualifications if approved by the agency.

Duties and Responsibilities of the Radiation Safety Committee (RSC)[43]

The RSC must perform the following duties/functions:

- Meet as often as necessary to conduct business (usually at least three times a year);
- Review summaries of information presented by the RSO, including over-exposures; significant incidents, such as spills, contamination, or medical events; and violations following an inspection;
- Review the program for maintaining doses ALARA, and providing any necessary recommendations to ensure doses are ALARA;
- Review the overall compliance status for authorized users;
- Share responsibility with the RSO to conduct periodic audits of the radiation safety program;
- Review the audit of the radiation safety program and act upon the findings;
- Develop criteria to evaluate training and experience of new authorized user applicants;
- Evaluate and approve authorized user applicants who request authorization to use radioactive material at the facility;
- Evaluate new uses of radioactive material;
- Review and approve permitted program and procedural changes prior to implementation; and
- Have knowledge of and ensure compliance with federal and state security measures for radioactive material.

Thus, the RSC actually controls all aspects of the broad license operations, with the RSO carrying out the day-to-day safety functions and reporting to the RSC.

---

[43] As presented in TAC §289.252(g)

4. License Application

Application for Specific License

Specific Licenses (Non-Broad Scope)

Each agency generally has prescribed license application forms. The forms collect uniform information about the applicant. Many states and the NRC also have regulatory guides available to assist in the application process. In general, a license application will consist of, as a minimum:

- A formal application form;
- Radiation safety procedures (such as a radiation safety manual or OSEP);
- Worker training procedures; and
- Other specific documentation required by the regulatory agency, such as a business information form identifying the applicant's incorporation or business status.

The information to be included in the application will probably include:

- Documentation showing that the applicant and all personnel who will be handling the radioactive material are qualified by reason of training and experience to use the material in question for the purpose requested as required by the rules in such a manner as to minimize danger to occupational and public health and safety and the environment;
- Documentation showing that The applicant's proposed equipment, facilities, and procedures are adequate to minimize danger to occupational and public health and safety and the environment;
- Evidence that the issuance of the license will not be inimical to the health and safety of the public;
- Evidence that the applicant satisfied any applicable special requirement in this rules;
- Documentation that the radiation safety information submitted for requested sealed source(s) or device(s) containing radioactive material is meets regulatory requirements;
- Documentation showing that the qualifications of the designated radiation safety officer (RSO) are adequate;
- An adequate operating, safety, and emergency procedures;
- Evidence that the applicant's permanent facility is located in the appropriate state (otherwise, if the applicant's permanent facility is not located in the state, reciprocal recognition will need to be sought to operate in that state);
- Documentation that the owner of the property is aware that radioactive material is stored and/or used on the property, if the proposed facility is

not owned by the applicant (not required for government owned property temporary job site use); and
- Business information identifying the applicant's incorporation or business status.

Application for Broad (Scope) License

The license applicant will be required to satisfy the requirements of the "specific license" listed above and also:

- Documentation that staff has substantial experience in the use of a variety of radioisotopes for a variety of research and development uses;
- Establishment of a full-time RSO meeting the requirements previously listed above; and
- Establishment of an RSC, including names and qualifications, with duties and responsibilities in accordance with subsection (g) of this section. The RSC shall be composed of an RSO, a representative of executive management, and one or more persons trained or experienced in the safe use of radioactive materials.

Not all activities using radioactive materials can be conducted under a broad license. Specific authorization or licensing is required for the following uses or activities:

- Tracer studies involving direct release of radioactive material to the environment;
- Receipt, acquisition, ownership, possession, use, or transfer of devices containing 100,000 curies (Ci) or more of radioactive material in sealed sources used for irradiation of materials;
- Performance or conduct of:
  - < commercial distribution of radioactive materials,
  - < manufacture of NARM devices,
  - < manufacture and commercial distribution of generally licensed devices,
  - < manufacture and commercial distribution of exempt sources.
- Addition, or cause the addition, of radioactive material to any food, beverage, cosmetic, drug, or other product designed for ingestion or inhalation by, or application to, a human being.

5. Registration Application

The difference between licensing and registration is rather simple. In general, licensing is required prior to taking possession of a radiation source (radioactive material) and registration is required after a radiation source (radiation producing machine, such as an X-ray machine) is already possessed. Registration may be of the x-ray device or of the use or of the user. All three, or combinations of the three, have been tried. Since only the states regulate x-ray device use (except for certain mammography situations) the regulatory approach will depend on the state in which the device will be used. The simplest approach is to register the device. In some cases it may be a system—for example a system with 2 or more x-ray heads and/or 2 or more control panels.

Registration Procedure

As in the licensing process, a prescribed application form is used to collect uniform information about the applicant. A rather simple OSEP is required and can often be supplied by a regulatory guide. Other application forms, such as a business information form, will probably be required to be completed and submitted with the application. In general, the application must be submitted within 30 days following the first use of the device. There are exceptions. Certain devices, such as accelerators and therapy devices may need to be registered in advance and may have more overall requirements. Since these devices offer a greater radiation threat, a regulatory agency will want to have more detail about safety systems before authorizing the device to be used. For example, It is better (and cheaper) to assure that a facility's shielding meets requirements before construction rather to have to retrofit it later.

6. Other Authorizations for Use of Radioactive Material

General License

As previously stated, there are two basic types of licenses: specific and general. The specific license requires advance application and agency approval, whereas the general license allows one to obtain a radioactive material device and put it use—followed by a form of registration. The authorization to use a device distributed as a generally licensed device is based in the radiation control rules. If a device has been evaluated by the US NRC, an AS, or a licensing state and approved to be added to the generally licensed device registry, then it can be manufactured and distributed as a generally licensed device. There are certain, specific criteria that must be met by the device—such as special labeling. Once listed, a person can then acquire the device and register its use after obtaining it. Basically, the general license is contained by the rules. However, the general license use by a licensee can be modified, suspended, and/or revoked if the regulatory agency finds violations and proceeds through its administrative procedures. Usually, generally licensed devices are passive devices, such as gauges and measuring devices affixed to plant "plumbing", as in a petrochemical plant. There is little human involvement, except for periodic inspection and leak testing of the sealed source. It should be noted that many devices, especially gauges, may be approved for distribution as both a specific or generally licensed device.

Exempt Radioactive Material

Certain radioactive materials can be distributed as exempt materials or sources. A person can be licensed to manufacture and distribute them. The end user is exempt from regulation. Even though they may be exempt, care should be taken in using, handling and storing them. For disposal, state requirements should be checked. Some examples of such devices/materials are:

- Timepieces, hands, or dials containing tritium or promethium-147 in µci and mci quantities.
- Lock illuminators containing tritium or promethium-147 in µci and mci quantities.
- Balances of precision containing tritium.
- Automobile shift quadrants containing tritium.
- Marine compasses and marine navigational instruments containing tritium.
- Thermostat dials and pointers containing tritium
- Electron tubes containing various isotopes in µci quantities.
- Ionizing radiation measuring instruments containing, for purposes of internal calibration or standardization, containing americium-241 in µci quantities.
- Spark gap irradiators containing not more than 1 µci of cobalt-60
- Ionization chamber smoke detectors containing not more than 1 microcurie (µci) of americium-241
- Self-luminous products containing tritium, krypton-85, promethium-147, or radium-226.
- Gas and aerosol detectors containing radioactive material capsules containing carbon-14 urea for "in vivo" diagnostic use in humans.

Often, these items become listed in the records of licensees with large inventories of materials and cause confusion during inspections

Reciprocity

A licensee in a state, licensed by the US NRC or by an AS, can use that license to conduct operations in another state, provided certain criteria are met. The criteria for Texas includes:

- the exact location, start date, duration, and type of activity to be conducted;

- the identification of the radioactive material to be used;
- the name(s) and in-state address(es) of the individual(s) performing the activity;
- a copy of the applicant's pertinent license;
- a copy of the licensee's operating, safety, and emergency procedures; and
- the appropriate fee.

The licensee must have an appropriate license and procedures and must provide 3 days notice to the state.

When contemplating working in another state, one should review the radiation control rules applicable in that state. It would be wise to call the state program, or NRC (both the NRC and the CRCPD have contact information online), and inquire about the process before making any commitments. If reciprocity is not obtained, the only other method of working in a state is to apply for a license for that state and receive the license before initiating work. You should note, however, that some states require a permanent office address in their state in order to obtain a license.

Temporary Job Sites

While the phrase "temporary job site" is used extensively throughout radiation control rules, the only place it is actually defined is in the SSRCR, Part E (industrial radiography). That definition is: "'Temporary jobsite" means a location where radiographic operations are performed and where sources of radiation may be stored other than the location(s) of use authorized on the license or registration."' The Texas rules also define temporary job site in its well-logging rules. This phrase should be formally defined in each state's rules, and those of the NRC, since differentiation between permanent and temporary sites can be an issue. A permanent (job) site is one that is listed on the license or application. Permanent sites are a continuing radiation threat to the area in which they are located and require that the regulatory agency fully evaluate that threat and assure that all safety necessary safety features are implemented by the licensee or registrant. Temporary sites may also present a radiation threat at a site, but the threat "comes and goes" and so a "missed hazard" should have much less of an effect over time. Special procedures are usually required at temporary job sites. For example, special surveys, temporary barriers, and written statements from the property owners will usually be required, as a minimum. The requirements are generally set up during the licensing process.

Termination of Sites

If a license or a site is terminated, one does not simply walk away from either one. Certain steps must be taken to end the process. First, advance notice is required. The agency must be notified and have opportunity to evaluate any site associated with the license or any permanent authorized site being removed from the license. The agency must assure that no residual radioactive material will remain at a site once the licensee relinquished control of the site. Even if the license expires, the licensee remains responsible for proper decommissioning of any licensed site.

Decommissioning

The Texas program requires: Within 60 days of the occurrence of any of the following, the licensee shall notify the agency in writing and either begin decommissioning its sites, buildings and/or areas that contain residual radioactivity, so that they are suitable for release in accordance the Texas rules or, within 12 months of notification, submit a decommissioning plan when required, and begin decommissioning upon approval of that plan if:

- the license has expired or has been revoked;
- the licensee has decided to permanently cease principal activities at the site, building, or area;
- no principal activities under the license have been conducted for a period of 24 months; or
- no principal activities have been conducted for a period of 24 months in each building or area that contains residual radioactivity (activities may be continuing at the site in other buildings or areas).

Other states will probably have similar requirements. It is important to carefully review those requirements, provide agency notification, properly dispose of radioactive materials, and properly release sites. Those that don't, suffer expensive consequences. It is important to work closely with the regulatory agency.

Chapter 7
Security for Radiation Sources

1. Introduction

With terrorist activities becoming more widespread throughout the world, most have become more concerned about the security of radioactive material. Many feel that the use of radioactivity in a terrorist device will be very much the worst the terrorists can do. While this is not true, we still need to provide the utmost security for radioactive materials. This chapter presents various security processes.

2. Basic Security

Early Security Requirements

Prior to the terror attacks of September 11, 2001 (9/11), security of radioactive materials was barely functional. For example, a user of a large industrial source might only be required to have a padlock on the storage container and a written procedure that stated "the material could only be handled by authorized person(s)". Radiation protection was the major concern of radiation regulatory organizations. After 9/11, the security requirements were greatly increased and solidified.

Current Security Requirements

Certain large quantities of radioactive material now have very stringent security requirements. These will be discussed in the next section. For the lesser quantities (less than Category 2), the requirements are simple and straightforward. Security must be addressed for radioactive materials in all phases: storage, use, and transport and any associated processes.

Regulatory Requirements for Security of Radioactive Material

For security and control of licensed sources of radiation, the radiation control rules require:

- The licensee shall secure radioactive material from unauthorized removal or access; and
- The licensee shall maintain constant surveillance, using devices and/or administrative procedures to prevent unauthorized access to use of radioactive material that is in an unrestricted area and that is not in storage.

Recently, the NRC added the following requirement:

> *Each portable gauge licensee shall use a minimum of two independent physical controls that form tangible barriers to secure portable gauges from unauthorized removal, whenever portable gauges are not under the control and constant surveillance of the licensee.*

The AS's are adding this requirement to their rules to remain compatible with the NRC rules. It is the RSO's responsibility to assure that these security features are implemented in the program(s) for which they are responsible.

Registered X-ray Device Security

As previously mentioned, the states regulate most uses of x-ray devices, usually requiring that they be registered. Since x-ray devices only operate when electric power is supplied, security is a simpler proposition. The radiation control rules generally, as a minimum, require that:

- radiation machines from be secured from unauthorized removal; and
- administrative procedures be used to prevent unauthorized use of radiation machines.

Some devices are equipped with key locking systems. Procedures establishing key use and removal should be established. There are also labeling requirements for radiation machines. Each registrant must ensure that each radiation machine is labeled in a "conspicuous manner that cautions individuals that radiation is produced when it is energized". The label(s) must be placed in a clearly visible location on the face of the control unit.

Security for therapy units and accelerators can have requirements more stringent than for other radiation sources, since they present high radiation levels. Controlled access during operation with interlock systems and audible warnings may be required. The RSO should carefully review the requirements in his/her state.

3. Advanced Security (Increased Controls)

Increased Controls

In 2005, the NRC issued an order requiring the development and implementation of "increased controls" by licensees (both NRC and AS licensees) possessing certain types and quantities of radioactive material. The requirements included more stringent procedures for allowing access to radioactive materials (such as documented background checks of authorized users) and implementation of security systems capable of initiating a timely armed response from a local law enforcement agency. In 2008, fingerprinting and an FBI background check were added to the requirements. At the same time, the AS issued orders, or amended licenses, requiring the same stringent controls.

Note: The Increased Controls (IC) information discussed in the following sections is readily available on the US NRC website and is, therefore, not sensitive or restricted.

Other NRC orders regarding security that have been issued:

- Panoramic and underwater irradiators
- Manufacturers and distributors
- Shipment of radioactive materials "Quantities of Concern" (RAMQC)
- Fingerprinting requirements for access to safeguards information
- Fingerprinting requirements for access to radioactive material
  - For irradiators
  - For manufacturers and distributors
- For RAMQC licensees
- Trustworthiness and reliability requirements for unescorted access to radioactive materials for service providers that are not manufacturers or distributors

If your program is involved in any of these areas, then you should review the area(s) on the NRC website to determine whether special requirements may affect your program.

The following is a summary paraphrased from the NRC Increased Controls Order. One should always defer to the written orders/rules/instructions of the radiation control agency with jurisdiction as this information can be incomplete, out of date, and/or changed by government action. This material is only intended to provide for familiarization of the requirements.

The quantities of concern are listed in the following table (Table 1 of the NRC initial IC order) downloaded from the NRC website:

Radionuclide	Quantity of Concern[a] (TBq)	Quantity of Concern[b] (Ci)
Am-241	0.6	16
Am-241/Be	0.6	16
Cf-252	0.2	5.4
Cm-244	0.5	14
Co-60	0.3	8.1
Cs-137	1	27
Gd-153	10	270
Ir-192	0.8	22
Pm-147	400	11,000
Pu-238	0.6	16
Pu-239/Be	0.6	16
Ra-226[c]	0.4	11
Se-75	2	54
Sr-90 (Y-90)	10	270
Tm-170	200	5,400
Yb-169	3	81
Combinations of radioactive materials listed above[d]	See Footnote Below	

Table 7-1 Quantities of Concern

Footnotes for Table 7-1:

[a] The aggregate activity of multiple, collocated sources of the same radionuclide should be included when the total activity equals or exceeds the quantity of concern.

[b] The primary values used for compliance with this order are TBq. The curie (Ci) values are rounded to two significant figures for informational purposes only.

[c] On August 31, 2005, the NRC issued a waiver, in accordance to Section 651(e) of the Energy Policy Act of 2005, for the continued use and/or regulatory authority of Naturally Occurring and Accelerator-Produced Material (NARM), which includes Ra-226. The NRC plans to terminate the waiver in phases, beginning November 30, 2007, and ending on August 7, 2009. The NRC has authority to regulate discrete sources of Ra-226, but has refrained from exercising that authority

[d] If several radionuclides are aggregated, the sum of the ratios of the activity of each source, i of radionuclide, n, A(i, n), to the quantity of concern for radionuclide n, Q(n), listed for that radionuclide equals or exceeds one. [(aggregated source activity for radionuclide A) ÷ (quantity of concern for radionuclide A)] + [(aggregated source activity for radionuclide B) ÷ (quantity of concern for radionuclide B)] + etc.....>1

In general, each licensee possessing radioactive material that meets the IC criteria of the NRC Increased Controls Order[44] must implement the following controls, as applicable:

Control Access to Radioactive Material Quantities of Concern (IC1)

If a licensee possesses radioactive material that are quantities of concern (RQC), or devices containing such materials, then the licensee must control access at all times to the materials/devices and must limit access to only approved individuals who require access to perform their duties. The licensee must assure that:

1. Only trustworthy and reliable individuals with licensee written approval have unescorted access to RQC.
2. Only those individuals with job duties that require access to RQC are approved for unescorted access.
3. Personnel who are not approved for unescorted access are escorted by an approved individual if they require access to RQC to perform a job duty.
4. Trustworthiness and reliability are determined, at a minimum, by verifying employment history, education, and personal references for individuals employed by the licensee for three years or less, and for non-licensee personnel, such as

[44] EA 05-090: Order Imposing Increased Controls, November 14, 2005.

physicians, physicists, house-keeping personnel, and security personnel under contract.

5. Independent information to corroborate that provided by the employee is obtained to the extent possible (i.e., seeking references not supplied by the individual).
6. The trustworthiness and reliability of Individuals employed by the licensee for longer than three years is determined, at a minimum, by a review of the employees' employment history with the licensee.
7. Fingerprinting procedures are implemented by June 2, 2008 (see fingerprinting in following section).
8. Service providers are escorted unless determined to be trustworthy and reliable by an NRC-required background investigation as an employee of a manufacturing and distribution (M&D) licensee, in which case written verification attesting to or certifying the person's trustworthiness and reliability is obtained from the manufacturing and distribution licensee providing the service.
9. The basis for concluding that there is reasonable assurance that an individual granted unescorted access is trustworthy and reliable, and does not constitute an unreasonable risk for unauthorized use of radioactive material quantities of concern is documented.
10. A list of persons approved for unescorted access to such radioactive material and devices is maintained.

Monitor and Detect Unauthorized Access (IC2)

If a licensee possesses radioactive material that are quantities of concern (RQC), or devices containing such materials, then the licensee must ensure the safe handling, use, and control of RQC material in use and in storage, must have a documented program to monitor and immediately detect, assess, and respond to unauthorized access to RQC. Enhanced monitoring must be provided during periods of source delivery or shipment, where the delivery or shipment exceeds 100 times the Table 1 values. The licensee must:

1. Respond immediately to any actual or attempted theft, sabotage, or diversion of such radioactive material or of the devices. The response shall include requesting assistance from a Local Law Enforcement Agency (LLEA).
2. Have a pre-arranged plan with LLEA for assistance in response to an actual or attempted theft, sabotage, or diversion of such radioactive material or of the devices which is consistent in scope and timing with a realistic potential vulnerability of the sources containing such radioactive material. The pre-arranged plan must be updated when changes to the facility design or operation affect the potential vulnerability of the sources. Prearranged LLEA coordination is not required for temporary job sites.

3. Have a dependable means to transmit information between, and among, the various components used to detect and identify an unauthorized intrusion, to inform the assessor, and to summon the appropriate responder.
4. As promptly as possible, notify the NRC Operations Center after initiating appropriate response to any actual or attempted theft, sabotage, or diversion of radioactive material or of the devices.
5. Maintain documentation describing each instance of unauthorized access and any necessary corrective actions to prevent future instances of unauthorized access.

Control Licensed Material During Transport (IC3)

If a licensee possesses and transports radioactive materials that are quantities of concern (RQC), or devices containing such materials, then the licensee must:

1. For domestic highway or rail shipments using a carrier other than the licensee, if the quantities equal or exceed those in Table 1 but are less than 100 times Table 1 quantities, per consignment, the licensee shall:
   a. Use carriers which:
      - Use package tracking systems,
      - Implement methods to assure trustworthiness and reliability of drivers,
      - Maintain constant control and/or surveillance during transit, and
      - Have the capability for immediate communication to summon appropriate response or assistance.
   b. The licensee shall verify and document that the carrier employs the measures listed above.
   c. Contact the recipient to coordinate the expected arrival time of the shipment;
   d. Confirm receipt of the shipment; and
   e. If the shipment does not arrive on or about the expected arrival time, initiate an investigation to determine the location of the licensed material. If it is determined the shipment has become lost, stolen, or missing, the licensee shall immediately notify the NRC Operations Center. If, after 24 hours of investigating, the location of the material still cannot be determined, the radioactive material shall be deemed missing and the licensee shall immediately notify the NRC Operations Center.
2. For domestic highway and rail shipments, prior to shipping licensed radioactive material that exceeds 100 times the quantities in Table 1 per consignment, the licensee shall:
   a. Notify the NRC, in writing, at least 90 days prior to the anticipated date of shipment. The NRC will issue the Order to implement the Additional Security Measures (ASMs) for the transportation of Radioactive Material Quantities of Concern (RAM QC). The licensee shall not ship this material until the ASMs

for the transportation of RAM QC are implemented or the licensee is notified otherwise, in writing, by NRC.

b. Once the licensee has implemented the ASMs for the transportation of RAM QC, the notification requirements shall not apply to future shipments of licensed radioactive material that exceeds 100 times the Table 1 quantities. The licensee shall implement the ASMs for the transportation of RAM QC.
c. If a licensee employs an M&D licensee to take possession at the licensee's location of the licensed radioactive material and ship it under its M&D license, the requirements of a. and b. above shall not apply.

3. If the licensee is to receive radioactive material greater than or equal to the Table 1 quantities, per consignment, the licensee shall coordinate with the originator to:
   a. Establish an expected time of delivery; and
   b. Confirm receipt of transferred radioactive material. If the material is not received at the expected time of delivery, notify the originator and assist in any investigation.

Physically Control Portable/Mobile Devices (IC4)

In order to ensure the safe handling, use, and control of licensed material in use and in storage each licensee that possesses mobile or portable devices containing radioactive material in quantities greater than or equal to Table 1 values, shall:

1. For portable devices, have two independent physical controls that form tangible barriers to secure the material from unauthorized removal when the device is not under direct control and constant surveillance by the licensee.
2. For mobile devices:
   a. That are only moved outside of the facility (e.g., on a trailer), have two independent physical controls that form tangible barriers to secure the material from unauthorized removal when the device is not under direct control and constant surveillance by the licensee.
   b. That are only moved inside a facility, have a physical control that forms a tangible barrier to secure the material from unauthorized movement or removal when the device is not under direct control and constant surveillance by the licensee.
3. For devices in or on a vehicle or trailer, licensees shall also use a method to disable the vehicle or trailer when not under direct control and constant surveillance by the licensee

Maintain Documentation of Controls (IC5)

Documentation required by increased controls must be maintained for three years after they are no longer effective:

1. Documentation regarding the trustworthiness and reliability of individual employees must be maintained for three years after the individual's employment ends.
2. Each time the list of approved persons or the documented program is revised, the previous documentation must be maintained for three years after the revision.
3. Documentation on each radioactive material carrier must be maintained for three years after the licensee discontinues use of that particular carrier.
4. Documentation on shipment coordination, notifications, and investigations must be maintained for three years after the shipment or investigation is completed.
5. Documentation required by increased controls must be maintained for three years after the license is terminated or amended to reduce possession limits below the quantities of concern.

Protect Sensitive Information from Unauthorized Disclosure (IC6)

Detailed information generated by the licensee that describes the physical protection of radioactive material quantities of concern, is sensitive information and must be protected from unauthorized disclosure. The licensee must:

1. Limit access to physical protection information to those persons who have an established need to know the information, and are considered to be trustworthy and reliable.
2. Develop, maintain and implement policies and procedures for controlling access to, and for proper handling and protection against unauthorized disclosure of, physical protection information for radioactive material covered by the orders and subsequent regulations. The policies and procedures shall include:
   a. General performance requirement that each person who produces, receives, or acquires the licensee's sensitive information, protect the information from unauthorized disclosure,
   b. Protection of sensitive information during use, storage, and transit,
   c. Preparation, identification or marking, and transmission,
   d. Access controls,
   e. Destruction of documents,
   f. Use of automatic data processing systems, and
   g. Removal from the licensee's sensitive information category.

Integrate Fingerprinting Procedures into Control of Access Procedures (Fingerprint Order[45])

Licensees who implemented a program to grant unescorted access to individuals due to possession of radioactive materials that are quantities of concern (RQC), or devices containing such materials, must modify their current trustworthiness and reliability program to include or address the following procedures:

1. Fingerprint each individual who is permitted unescorted access to risk significant radioactive materials, and have established a need to access any such material, equal to or greater than the RQC quantities. The Licensee must review and use the information received from the Federal Bureau of Investigation (FBI) identification and criminal history records check and ensure that the criteria are satisfied.
2. Notify each affected individual that the fingerprints will be used to secure a review of his/her criminal history record and inform the individual of the procedures for revising the record or including an explanation in the record.
3. Fingerprints for unescorted access need not be taken if:
   a. An employed individual (e.g., a Licensee employee, contractor, manufacturer, or supplier) is relieved from the fingerprinting requirement by 10 CFR § 73.61, or any person who has been favorably-decided by a U.S. Government program involving fingerprinting and an FBI identification and criminal history records check within the last five (5) calendar years, or any person who has an active federal security clearance (provided in the latter two cases that they make available the appropriate documentation.
   b. Written confirmation from the Agency/employer which granted the federal security clearance or reviewed the FBI criminal history records results based upon a fingerprint identification check is provided. The Licensee must retain this documentation for a period of three (3) years from the date the individual no longer requires unescorted access to certain radioactive material associated with the Licensee's activities.
4. All fingerprints obtained by the Licensee pursuant to this Order must be submitted to the Commission for transmission to the FBI. Additionally, the Licensee shall submit a certification of the trustworthiness and reliability of the T&R Official.
5. The Licensee must review the information received from the FBI and consider it, in conjunction with the trustworthiness and reliability requirements of the IC Order (EA-05-090), in making a determination whether to grant unescorted access to certain radioactive materials.

45 NRC Order EA-07-305: Issuance of Order Imposing Fingerprinting and Criminal History Records Check Requirements for Unescorted Access to Certain Radioactive Material, December 5, 2007.

6. The Licensee must use any information obtained as part of a criminal history records check solely for the purpose of determining an individual's suitability for unescorted access to risk significant radioactive materials equal to or greater than the RQC quantities.
7. The Licensee must document the basis for its determination whether to grant, or continue to allow unescorted access to risk significant radioactive materials equal to or greater than the quantities listed in attachment 2.

Prohibitions

1. A Licensee shall not base a final determination to deny an individual unescorted access to certain radioactive material solely on the basis of information received from the FBI involving: an arrest more than one (1) year old for which there is no information of the disposition of the case, or an arrest that resulted in dismissal of the charge or an acquittal.

2. A Licensee shall not use information received from a criminal history check obtained pursuant to this Order in a manner that would infringe upon the rights of any individual under the First Amendment to the Constitution of the United States, nor shall the Licensee use the information in any way which would discriminate among individuals on the basis of race, religion, national origin, sex, or age.

Caution: One should refer carefully to the NRC orders when setting up their IC program. The steps discussed here are introductory and not all inclusive.

4. Enhanced Security

Over the years, security of radioactive material has been improved in its effectiveness out of necessity. It has gone from simple padlocking to the stringency of the increased controls previously discussed. However, not all materials fall under increased controls and many of those that do not could be used by ill intentioned persons in attempts to harm or terrorize others. This can be done by using an accumulation of stolen materials in place of a large source. Or, it could be done by simply using a small quantity, say in an explosion and then claiming it's a larger quantity. If a person set off a radioactive dispersal device (RDD) containing 100 millicuries of Co-60 and claimed that it contained 100 Ci (an IC quantity), it would probably take days to discover the true quantity. This would produce the effect of days of terror. Thus, it is felt by many that security for most radioactive materials should be greatly improved to help preclude this type of nefarious use.

Enhanced Security Program(ESP)

An "enhanced security program" for radioactive materials has been developed by a research team. In 2009, the Federal Bureau of Investigation (FBI) and Interpol expressed their concern about the potential use of radioactive material as a terrorist tool, due to inadequate security, to ASME and the Alfred P. Sloan Foundation. Consequently, ASME Innovative Technologies Institute, LLC (ASME-ITI) applied for and received a grant from the Sloan Foundation to develop a risk-based methodology to help identify and prioritize significant risk to the public from radioactive materials used in the medical, industrial, and academic communities (MIAN). The author was privileged to serve on the research team.

The outcome of the research produced, among other security information and tools, the "**Enhanced Security Program For MIAN Facilities** (09/10/2012)", which is available for no charge to licensees. While a portion of the program is presented here, full information, as well as a screening tool, can be obtained by visiting the website ***SecureRAM.com***.

The Enhanced Security Program (ESP) is a proactive program consisting of a screening process that enables possessors of radioactive material to voluntarily assess their own level of security and make modifications to enhance security to higher levels, if necessary.

The steps of ESP consist of:

1. Using a screening tool to assess the current security level.
2. Using a screening/assessment tool to determine additional processes that could be added to enhance security.
3. Employing additional security measures in the facility's security plan to reduce the potential risk for the material being used as a weapon.
4. Repeating Steps 1-3 until security is increased to an acceptable level.

ESP considers and includes the use of Increased Controls for Category I and 2 Radioactive Materials required by government regulations. It also includes Category 3 and 4 radioactive materials, as well as Category 5 sources, although the rather innocuous Category 5 sources need not be addressed.

When dealing with the protection of assets (radioactive material in this case), there are basically two processes that can be established. One option is to set up a physical protection system (PPS) that will absolutely prohibit the unauthorized removal of radioactive material. A second option is to set up a PPS that will detect intrusion and delay the intruder until an appropriate response can be initiated. Since the first option is virtually impossible, the second is the approach adopted by ESP.

The security "philosophy" embraced herein follows, in part, along the lines of security procedures and processes developed at Sandia Laboratories. In most cases involving significant quantities of radioactive material, a PPS should be incorporated to provide protection during use, storage, and transport of the material. During use and transport, workers are usually present and security issues are often less of a problem. Thus, ESP will deal primarily with the condition of storage of radioactive materials.

Like increased controls, ESP incorporates a system of physical barriers, detection systems, and adequate response in the event of intrusion. Physical barriers are objects that prevent unauthorized persons from having access to the radioactive material. The simplest barrier is a locked container holding the radioactive material. Other barriers might include a locked room and/or building, or an entry controlled fence. Obviously, the stronger the physical barrier the better it is as a deterrent. A fifteen foot tall, one-foot thick concrete wall is probably a much better deterrent/barrier than a four foot tall chain link fence. However, rather than line up and measure the value of various types of walls, ESP assumes that the established barrier is "adequate". ESP does not evaluate the various detection systems. Again, it assumes that the installed system will be adequate.

While physical barriers are generally a requirement for entities possessing radioactive material, detection systems are not—nor is there a requirement for local law enforcement agency (LLEA) armed response for all radioactive material possessors. ESP proposes, however, using the best of the security methods available and practicable for each possessor/licensee. By evaluating the security level for each source of radioactive material, or for each collection of sources, we can make an assessment to determine whether the level of security is adequate for that particular material. If it is found to be inadequate, then additional security measures can be incorporated to increase the security level to a more adequate one. In the many cases where there is no regulatory requirement for detection systems, the possessor/licensee may determine that his/her security level needs to be increased and may opt to add a detection system to accomplish that. If the possessor/licensee endeavors to increase the security level by adding a detection system, then the "plan" should be to assure that the physical barriers are sufficient to delay an intruder long enough so that an appropriate response can be accomplished before the intruder can depart the site with the radioactive material.

ESP addresses "levels of security", where one level of security is a device and/or method designed to prevent access. For example, a source stored in a locked container would be one level of security. Some examples of individual methods providing one level of security are:

- Locked device or container
- Device/container chained or positively secured to structure, or is physically part of structure
- Locked door to area
- Locked building.
- Locked fence around site.
- Video/audio surveillance.
- Guard.
- Alarmed/monitored security system.
- Presence of authorized personnel.

Thus, placing a locked container in a locked room, in a locked building, inside of a locked fenced area, with a security guard on duty 24 hours per day, would provide a total of 5 levels of security—if the value of each method is 1. Personnel in attendance would be one level, as would an operable intruder alert system. On the other hand, a system of background checks would not be counted, but should be considered under the overall security plan of the facility. Not all locked features would be counted as one security level. If a room was used to store radioactive material and the room had two (2) locked doors, only one security level exists. Either door could be penetrated so there is only one level of security. It is also anticipated that the locks being used are effective and control of keys and codes must be considered.

It is anticipated that radioactive material could be removed from the possession of a licensee by one of the following:

- Diversion of shipment to/from site
- Inside employee removes
- Theft
- Armed attack team

Ideally, a certain level of security could be set to prevent any of the above methods from being successful. Unfortunately, almost all facilities are different and a given level which is successful for one facility may be woefully inadequate for the next. Thus, each facility should be evaluated on its own merits with the security levels being used as guides.

There are many methods of deploying any stolen radioactive material, but the most likely methods would be drawn from:

- Explosive dispersal of sealed sources
- Placement of individual or clustered sources in transportation system or public areas, such as schools, universities, and government offices, to cause exposure of individuals
- Radioactive material removed from cladding and placed in dispersible condition—dispersed by explosion
- Radioactive material removed from cladding and placed in dispersible (water soluble) condition with dispersion into water or food supply (schools, universities, commercial businesses, public venues, and government offices could be targeted)
- Perpetrators hide material in unknown location and use fear of exposure to terrorize the public

The consequences and costs of the deployment of radioactive material through one of these mechanisms should be evaluated so that the efforts of preventing removal and deployment will be expended in the areas with greatest consequence.

For storage, use, and/or transport circumstances, the selected security methods from the following lists should be employed in a manner that renders unauthorized removal to be highly unlikely.

Some security methods that can be used are:

Active Methods:

- Periodic inspection/inventory.
- Crisis Response and Recovery Plans
- Presence of authorized personnel.
- X-ray of incoming/outgoing packages
- Metal detection systems
- System of authorizing access (references, but no background checks).
- Continuing/periodic security training.
- Public Service Liaison
- Vendor and Contractor Access Control
- System of authorizing access (background checks).

- Controls limiting access according to regulations/rules
- Video/audio surveillance.
- Daily inspection/inventory.
- Card Access Control
- Key Control Policy
- Key System Manager
- Incident Reporting Methods and Facilitation
- Incident Response Mechanism
- Rapid, dependable means of communication (cell phones, radios, etc).
- Identification and Badging System
- Guard (unarmed).
- 24/7 continuous surveillance by operator personnel
- Theft and Loss Prevention—Shipping & Receiving
- Method to deploy armed local law enforcement (LLEA).
- Remote monitoring of CCTV or assessment by operator
- Physical searches of persons at access points.
- Procedures to identify and protect sensitive information
- Security training for personnel.
- Security plan/program.
- Armed guard

Passive Methods:

- Locked device.
- Locked container.
- Container chained or positively secured to structure.
- Locked metal (steel) cage.
- Locked door to room of use or storage.
- Locked fence around site without entrance control
- Exterior Security Lighting
- Locked building.
- Locked fence around site with entrance control.
- Device/container physically part of structure.
- Device position in structure (such as elevated on platform).
- Monitoring/alarm system.
- External Barriers (Fences, Gates, Etc.), Perimeter Control, Interior Barriers
- Electronic tamper detection equipment
- Access Control and Alarm Monitoring System (ACAMS)—Hardware, software
- Electronic Locking Devices
- Exterior Area Intrusion Detection

Each licensee (a possessor of radioactive material must have a license issued by the US NRC or an Agreement State) should employ a number of these security methods to minimize the possibility of unauthorized removal. A licensee under increased controls, for example, would probably implement the following, as a minimum:

- Security plan/program.
- Security training for personnel.
- System of authorizing access.
- Monitoring/alarm system.
- Method to deploy armed local law enforcement (LLEA).
- Periodic inspection/inventory.
- Locked device/container.
- Container chained or positively secured to structure.
- Locked door to room of use or storage.
- Locked building.
- Locked fence around site.

Assuming that the value of each security level is one (1), this would yield a security level of about eleven (11). A licensee NOT under increased controls but possessing a gauging device with a very large source would probably employ the following, as a minimum:

- Security plan/program or a system of authorizing access.
- Periodic inspection/inventory.
- Locked device/container.
- Container chained or positively secured to structure.
- Locked door to room of use or storage.
- Locked fence around site.

This would only be a security level of about six (6) or seven (7). In most cases, experience indicates a security level of 3 or 4 is typical for non-IC licensees.

This program can be improved by using a weighted scheme of security levels. Some security measures are often more valuable/effective than others in prevention, or at least slowing down, unauthorized access. In most circumstances, an armed security guard would be far more effective than a padlock—even a heavy duty one. Thus, weighting of the security devices/methods will be employed by ESP to better estimate the overall security level. The weighting of the various methods will be presented in the screening tool.

IAEA Security Levels

The IAEA has presented a system of categorizing radioactive material according to its relative danger[46] and setting security levels according to the category. The highest category on this systems is Category I (most dangerous) and the lowest is Category 5. IAEA established Security Levels A, B, and C with A having the most stringent security measures. Application of the Security Levels is as follows:

Category	Security Level	Purpose
Category I	Security Level A	Prevent unauthorized removal of a source
Category II	Security Level B	Minimize the likelihood of unauthorized removal of a source
Category III	Security Level C	Reduce the likelihood of unauthorized removal of a source.

Categories IV & V should use the "International Basic Safety Standards" given in the "International basic safety standards[47] for protection against ionizing radiation and for the safety of radiation sources" publication. For convenience, ESP has set up comparative tables so that the IAEA security procedures can be used to set minimum security levels. Tables are also set for Categories 4 & 5, although the regulatory consensus is that more stringent controls for Category 5 are generally unnecessary.

ESP Screening Tool

ESP was modified to use a screening tool that has been developed to assess the security level for each source, to assess the risk that it might be used as a weapon, and to assess the consequences and cost should the source be successfully deployed as a weapon. The licensee should apply the screening tool to each source possessed that exceeds one (1) IAEA D-level. If the screening process indicates that the security level is too low and/or that the risk of it being used as a weapon and/or that the cost/ consequences would be unacceptably high, then additional security steps may need to be employed. The risk assessment methodology for MIAN sites is contained in the

46 Security of Radioactive Sources Implementing Guide International Atomic Energy Agency Vienna, 2009 IAEA Nuclear Security Series No. 11, P. 28

47 International Basic Safety Standards for Protection Against Ionizing Radiation and for the Safety of Radiation Sources.—Vienna: International Atomic Energy Agency, 1996, p.; 24 cm.—(Safety series, ISSN 0074-1892; 115. Safety Standards) STI/PUB/996 ISBN 92-0-104295-7.

two documents: "[48]Methodology for Assessing Risk from Radioactive Materials Found in Medical, Industrial and Academic Sites", March 2011; and "[49]The MIAN Security Enhancement Process", January, 2013. Additional information can be obtained, including access to the ESP, at the website:

SecureRAM.com

Site security has a large impact on the overall risk to the public regarding the use of radioactive material for terrorist purposes. One of the key parameters that will reduce overall risk is to reduce the probability of obtaining the material for malicious purposes. While it is impossible to completely eliminate the possibility of theft, armed attack, and other extreme means of obtaining material at all sites, good security practices can greatly reduce the probability of success by the terrorist or other criminal.

---

[48] Methodology for Assessing Risk from Radioactive Materials Found in Medical, Industrial and Academic Sites", March 2011. J. William Jones, PhD, Robert E. Nickell, PhD, John Haygood, MS. ASME Innovative Technologies Institute, LLC, Washington DC. Funded by the Sloan Foundation, NY, NY.

[49] The MIAN Security Enhancement Process", January, 2013. J. William Jones, PhD, Robert E. Nickell, PhD, John Haygood, MS. J. William Jones Consulting Engineers, Inc., Huntington Beach, Ca. Funded by the Sloan Foundation, NY, NY.

Appendix 1

List of Training Topics

The following topical areas of training were consolidated from the various requirements in radiation control rules and regulatory guides. A training program can be developed by choosing the topics applicable to your use of radiation sources.

I. Fundamentals Of Radiation Safety
   A. Characteristics of radiation
   B. Units of radiation dose (rem) and quantity of radioactivity (curie)
   C. Significance of radiation dose
      1. Radiation protection standards
      2. Biological effects of radiation dose
      3. Case histories of accidents
   D. Levels of radiation from sources of radiation
   E. Methods of controlling radiation dose
      1. Working time
      2. Working distances
      3. Shielding
      4. ALARA philosophy
   F. Radiation safety practices
      1. Prevention of contamination
      2. Methods of decontamination
   G. Discussion of ingestion, inhalation pathways
   H. Sources of Radiation
      1. natural
      2. man-made
      3. NORM at the job site
   I. Biological Effects of Radiation
      1. stochastic effects
      2. non-stochastic effects

II. Radiation Detection Instrumentation To Be Used
   A. Use of radiation survey instruments
      1. Operation
      2. Calibration
      3. Limitations
      4. types, probes
   B. Survey techniques
      1. equipment
      2. land
      3. buildings
      4. regulatory limits
   C. Use of personnel monitoring equipment (dose assessment)
      1. Film badges
      2. Thermoluminescent dosimeters (TLDs)
      3. Pocket dosimeters
      4. Alarming ratemeters
   D. Equipment to be used
      1. Handling equipment and remote handling tools
      2. Sources of radiation
      3. Storage, control, disposal, and transport of equipment and sources of radiation
      4. Operation and control of equipment
      5. Maintenance of equipment
   E. Internal Dose Assessment
      1. air monitoring
      2. bioassays
      3. whole body counting
      4. calculation of total effective dose equivalent (TEDE)
   F. Measuring,/Analytical Equipment
      1. Gamma spectroscopy
      2. Alpha/beta analysis
      3. Radon analysis
   G. Sampling/Monitoring Techniques
      1. air
      2. soil
      3. water
      4. radiation levels
   H. Analytical Methodologies and Equipment
      1. air samples
      2. soil samples
      3. water samples

III. Requirements Of Pertinent Federal And State Regulations
  A. Posting of warning signs
  B. Warning, Content, and Transport Labels
  C. Notices to workers
  D. License requirements and restrictions
  E. Workers' rights
  F. Licensee's Procedures (written operating and safety procedures)
    1. instrument calibration
    2. receipt, transfer and disposal
    3. dose assessment
    4. dose to general public
    5. surveys
    6. analytical results
    7. radiation protection program review
    8. worker training
    9. records maintenance.

IV. Generic Written Operating, Safety, And Emergency Procedures

V. Equipment To Be Used
  A. Remote handling equipment, tools
  B. Radiation sources
    1. Operation and control of radiographic exposure devices and sealed sources, including pictures or models of source assemblies (pigtails)
    2. Sources of radiation (logging tools, tracers, etc.)
  C. Storage and transport containers, source changers
  D. Operation and control of equipment
  E Collimators
  F. Equipment maintenance

VI. Records
  A. record keeping procedures
  B. Records
    1. instrument calibration
    2. receipt, transfer and disposal
    3. dose assessment
    4. dose to general public
    5. surveys
    6. analytical results
    7. radiation protection program review
    8. worker training

VII. Tracer Study Procedures
  A. Sources of contamination;

B. Contamination detection and control;
C. Decontamination techniques and limits;
D. Survey techniques for tracer materials;
E. Packaging requirements for transportation of radioactive materials—especially residual materials from tracer studies.

Appendix 2

Table of Information for Selected Common Radioisotopes Used for Sealed Sources

Appendix 2—Table of Information for Selected Common Radioisotopes Used for Sealed Sources

Isotope	Half-life	Radiation Levels R/hr @ distance per curie	Principal Radiations Types & Energies	Parent/Source & Daughters	Container Label Quantity	CRM Posting Quantity
Key	**a—alpha b—beta d—day dr—daughter f—foot g—gamma/x-ray m—meter n—neutron nc- neutron capture p—parent/source y—year n (Be)—neutrons produced by Beryllium atom interactions**					
Americium-241 **($_{95}$Am^{241})**	458 y		^{a} 5.49 MeV ^{g} 60 KeV n (Be)	^{p} Pu-241 (nc) ^{d} Pu-241	0.0001 µCi 3.7 Bq	0.001 µCi 37 Bq
Californium-252 ($_{98}$**Cf**^{252})	2.646 y		^{a} 6.12 MeV ^{n} (fission - high energy)	^{p} U-238 (nc) ^{p} Pu-239 (nc) ^{d} Cm-248	0.001 uCi 37 Bq	0.01 µCi 370 Bq
Cesium-137 (**Cs**^{137})	30.0 y	^{g} 0.33 @ 1 m ^{g} 3.65 @ 1f	^{g} 662 KeV	^{p} Fission Product ^{d} Xe-137, ^{d} Ba-137m	10.0 µCi 0.37 mBq	100.0 µCi 3.7 mBq
Cobalt-57 ($_{27}$**Co**^{57})	270 d	^{g} 0.09 @ 1 m ^{g} 0.97 @ 1f	^{g} 122 KeV	^{p} Ni-58 ^{p} Mn-55 ^{p} Fe-56 ^{d} Ni-57	100.0 µCi 3.7 mBq	1000.0 µCi 37 mBq
Cobalt-60 (^{27}**Co**^{60})	5.26 y	^{g} 1.32 @ 1 m ^{g} 14.2 @ 1f	^{g} 1.173 MeV ^{g} 1.332 MeV	^{p} Co-59 ^{d} Stable cobalt	1.0 µCi 37 kBq	10.0 µCi 0.37 mBq
Indium-114m ($_{49}$**In**^{114m})	50.0 d	^{g} 0.02 @ 1 m ^{g} 0.22 @ 1f	^{g} 192 KeV ^{g} 724 KeV ^{g} 558 KeV	^{p} In-113 ^{d} In-114	10.0 µCi 0.37 mBq	100.0 µCi 3.7 mBq
Iodine-125 (^{53}**I**^{125})	60.2 d	^{g} 0.07 @ 1 m ^{g} 0.75 @ 1f	^{g} 35 KeV	^{p} Sb-123 ^{d} Xe-125	1.0 µCi 37 kBq	10.0 µCi 0.37 mBq

Appendix 2—Table of Information for Selected Common Radioisotopes Used for Sealed Sources

Isotope	Half-life	Radiation Levels R/hr @ distance per curie	Principal Radiations Types & Energies	Parent/Source & Daughters	Container Label Quantity	CRM Posting Quantity
Key	**a—alpha b—beta d—day dr—daughter f—foot g—gamma/x-ray m—meter n—neutron nc- neutron capture p—parent/source y—year n (Be)—neutrons produced by Beryllium atom interactions**					
Iridium-192 ($_{77}$**Ir** ^{192})	74.2 d	^{g} 0.48 @ 1 m ^{g} 5.20 @ 1f	^{g} 317 KeV ^{g} 468 KeV ^{g} 308 KeV ^{g} 296 KeV	^{p} Ir-191 (nc) ^{d} Ir-192m	10.0 µCi 0.37 mBq	100.0 µCi 3.7 mBq
Plutonium-239 ($_{94}$**Pu** ^{239})	24,390 y		n (Be) 10.6 MeV max, 3.5 MeV avg	^{p} U-238, U-239 ^{d} Np-239	0.001 uCi 37 Bq	0.01 µCi 370 Bq
Radium-226 ($_{88}$**Ra** ^{226})	1602 y	^{g} 0.51 @ 1 m ^{g} 5.50 @ 1f	^{g} 186 KeV ^{n} (Be) 13 MeV max, 5 MeV avg	^{p} Th-230 ^{d} Rn-222	0.1 µCi 3.7 kBq	1.0 µCi 37 kBq

CAUTION—The above information has been compiled from various resources for convenience and demonstration. The accuracy or acceptability of any of the values is not guaranteed. Further, information taken from the Texas Regulations for Control of Radiation is subject to change at any time. One should always refer to current, or the most recent, accepted values when the performing calculations or making determinations where results impact health and safety and/or regulatory compliance.

Appendix 3

Glossary of Radiation Terms

Downloaded from the radiation control website of the Texas Department of State Health Resources, and reformatted.

Absorption The process by which the number of particles or photons entering a body of matter is reduced by interaction with matter. Also the process in which energy is absorbed from the particles or photons even if the number is not reduced.

Air Sampling The collection of samples of air to measure the radioactivity or to detect the presence of radioactive material, particulate matter, or chemical pollutants in the air.

ALARA As Low As Reasonably Achievable. This means making every reasonable effort to maintain exposures to radiation as far below the dose limits as practical consistent with the purpose for which the licensed activity is undertaken, taking into account the state of technology, the economics of improvements in relation to state of technology, the economics of improvements in relation to benefits to the public health and safety, and other societal and socioeconomic considerations, and in relation to utilization of nuclear energy and licensed materials in the public interest.

Alpha A positively charged particle ejected spontaneously from the nuclei of some radioactive elements. It has low penetrating power and a short range. The most energetic alpha particle will generally fail to penetrate the dead layers of cells covering the skin. Alphas are hazardous when an alpha-emitting isotope is inside the body.

Atom The smallest particle of an element that cannot be divided or broken up by chemical means. It consists of a central core of protons and neutrons, called the nucleus. Electrons revolve in orbits in the region surrounding the nucleus.

Atomic Energy Commission (AEC) Federal agency created in 1946 to manage the development, use, and control of nuclear energy for military and civilian application. Abolished by the Energy Reorganization Act of 1974 and succeeded by the Energy Research and Development Administration (now part of the U.S. Department of Energy) and the U.S. Nuclear Regulatory Commission.

Background Radiation Radiation from cosmic sources, naturally occurring radioactive materials, including radon (except as a decay product of source or special nuclear material) and global fallout as it exists in the environment from testing of nuclear explosive devices. It does not include radiation from source, byproduct, or special nuclear materials regulated by the NRC. The typically quoted average individual exposure from background radiation is 360 millirems per year.

Becquerel (Bq) The unit of radioactive decay equal to 1 disintegration per second. 3.7 X 10^{10} Bq = 1 Curie.

Beta A charged particle emitted from a nucleus during radioactive decay, with a mass equal to 1/1837 that of a proton. A negatively charged beta particle is identical to an electron. A positively charged beta particle is called a positron. Large amounts of beta radiation may cause skin burns, and beta emitters are harmful if they enter the body. Beta particles may be stopped by thin sheets of metal or plastic.

Bioassay The determination of kinds, quantities or concentrations, and in some cases, the locations of radioactive material in the human body, whether by direct measurement (in vivo counting) or by analysis and evaluation of materials excreted or removed from the human body.

Calibration The adjustment, as necessary, of a measuring device such that it responds within the required range and accuracy to known values of input.

Charged Particle An ion. an elementary particle carrying a positive or negative electric charge.

Contamination The deposition of unwanted radioactive material on the surfaces of structures, areas, objects, or personnel. It may also be airborne or internal (inside components or personnel).

Cosmic Radiation Penetrating ionizing radiation, both particulate and electromagnetic, originating in outer space. Secondary cosmic rays, formed by interactions in the earth's atmosphere, account for about 45 to 50 millirem of the 360 millirem background radiation that an average individual receives in a year.

Counter A general designation applied to radiation detection instruments or survey meters that detect and measure radiation. The signal that announces an ionization event is called a count.

CRCPD Conference of Radiation Control Program Directors

Curie (Ci) The basic unit used to describe the intensity of radioactivity in a sample of material. The curie is equal to 37 billion disintegrations per second, which is approximately the rate of decay of 1 gram of radium. A curie Is also a quantity of any radionuclide that decays at a rate of 37 billion disintegrations per second. Named for Marie and Pierre Curie, who discovered radium in 1898.

Daughter Products Isotopes that are formed by the radioactive decay of some other isotope. In the case of radium 226, for example, there are 10 successive daughter products, ending in the stable isotope lead -206.

Decay, Radioactive The decrease in the amount of any radioactive material with the passage of time, due to the spontaneous emission from the atomic nuclei of either alpha or beta particles, often accompanied by gamma radiation.

Declared Pregnant Woman A woman who has voluntarily informed her employer, in writing, of her pregnancy and the estimated date of conception.

Decontamination The reduction or removal of contaminating radioactive material from a structure, area, object, or person. Decontamination may be accomplished by:

1) Treating the surface to remove or decrease the contamination.

2) Letting the material stand so that the radioactivity is decreased as a result of natural decay.

3) Covering the contamination to shield or attenuate the radiation emitted.

Detector A material or device that is sensitive to radiation and can produce a response signal suitable for measurement or analysis.

DOE (United States) Department of Energy

Dose, Absorbed The amount of energy deposited in any substance by ionizing radiation per unit mass of the substance. It is expressed numerically in rads or grays.

Dose Equivalent A term used to express the amount of biologically effective radiation dose when modifying factors have been considered. The product of absorbed dose multiplied by a quality factor multiplied by a distribution factor. It is expressed numerically in rems or sieverts.

Dose Rate The radiation dose delivered per unit time. For example, rem per hour.

Dose The absorbed dose, given in rads or grays, that represents the energy absorbed from the radiation in a gram of any material. Furthermore, the biological dose or dose equivalent, given in rem or sieverts, is a measure of the biological damage to living tissue from the radiation exposure.

Dosimeter A portable instrument for measuring and registering the total accumulated dose to ionizing radiation.

Dosimetry The theory and application of the principles and techniques involved in the measurement and recording of radiation doses.

DOT (United States) Department of Transportation

Effective Half life The time required for the amount of a radioactive element deposited in a living organism to be diminished 50% as a result of the combined action of radioactive decay and biological elimination.

Electromagnetic Radiation	A traveling wave motion resulting from changing electric or magnetic fields. Familiar electromagnetic radiation range from X rays (and gamma rays) of short wavelength, through the ultraviolet, visible, and infrared regions, to radar and radio waves of relatively long wavelength. All electromagnetic radiations travel in a vacuum with the velocity of light.
Electron	An elementary particle with a negative charge and a mass 1/1837 that of the proton. Electrons surround the positively charged nucleus and determine the chemical properties of the atom.
Element	One of the 103 known chemical substances that cannot be broken down further without changing its chemical properties. Some examples include hydrogen, nitrogen, gold, lead, and uranium.
EMF/ELF	Electric and Magnetic Fields
Exposure	Being exposed to ionizing radiation or to radioactive material.
External Radiation	Exposure to ionizing radiation when the radiation source is located outside the body.
Extremities	The hands, forearms, elbows, feet, knee, leg below the knee, and ankles. (Permissible radiation exposures in these regions are generally greater than the whole body because they contain less blood forming organs and have smaller volumes for energy absorption.)
FDA	(United States) Food and Drug Administration
FEMA	Federal Emergency Management Administration
Film Badge	A pack of photographic film used for measurement of radiation exposure for personnel monitoring purposes. The badge may contain two or three films of differing sensitivities, and it may contain a filter that shields part of the film from certain types of radiation.
Gamma Ray	High energy, short wavelength, electromagnetic radiation (a packet of energy) emitted from the nucleus. Gamma radiation frequently accompanies alpha and beta emissions and always accompanies fission. Gamma rays are very penetrating and are best stopped or shielded by dense materials, such as lead or uranium. Gamma rays are similar to X-rays.

Geiger Mueller Counter A radiation detection and measuring instrument. It consists of a gas-filled tube containing electrodes, between which there is an electrical voltage, but no current flowing. When ionizing radiation passes through the tube, a short, intense pulse of current passes from the negative electrode to the positive electrode and is measured or counted. The number of pulses per second measures the intensity of the radiation field. It was named for Hans Geiger and W. Mueller, who invented it in the 1920's. It is sometimes called simply a Geiger counter or a G M counter.

Half life The time in which one half of the atoms of a particular radioactive substance disintegrates into another nuclear form. Measured half-lives vary from millionths of a second to billions of years. Also called physical or radiological half-life.

Half life, Biological The time required for the body to eliminate one half of the material taken in by natural biological means.

Half value, layer The thickness of any given absorber that will reduce the intensity of a beam of radiation to one half of its initial value.

Health Physics The science concerned with recognition, evaluation, and control of health hazards from ionizing radiation.

Health Physicist A person who works in the area of environmental health engineering that deals with the protection of the individual and population groups against the harmful effects of ionizing and non-ionizing radiation. The health physicist is responsible for the safety aspects in the design of processes, equipment, and facilities utilizing radiation sources, so that radiation exposure to personnel will be minimized, and will at all times be within acceptable limits.

High Radiation Area Any area with dose rates greater than 100 mrem in one hour 30 cm from the source or from any surface through which the radiation penetrates. These areas must be posted as "high radiation areas" and access into these areas is maintained under strict control.

Hot A colloquial term meaning highly radioactive.

Hot Spot The region in a radiation/contamination area in which the level of radiation/contamination is noticeably greater than in neighboring regions in the area.

HPS Health Physics Society.

Internal Radiation Nuclear radiation resulting from radioactive substances in the body. Some examples are iodine-131 (found in the thyroid gland) and strontium-90 and plutonium-239 (found in bone).

Ion 1) An atom that has too many or too few electrons, causing it to have an electrical charge, and therefore, be chemically active.

2) An electron that is not associated (in orbit) with a nucleus.

Ionization The process of adding one or more electrons to, or removing one or more electrons from, atoms or molecules, thereby creating ions. High temperatures, electrical discharges, or nuclear radiations can cause ionization.

Ionization Chamber An instrument that detects and measures ionizing radiation by measuring the electrical current that flows when radiation ionizes gas in a chamber, making the gas a conductor of electricity.

Ionizing Radiation Any radiation capable of displacing electrons from atoms or molecules, thereby producing ions. Some examples are alpha, beta, gamma, X-rays, neutrons. High doses of ionizing radiation may produce severe skin or tissue damage.

Irradiation Exposure to radiation.

Isotope One of two or more atoms with the same number of protons, but different numbers of neutrons in their nuclei. Thus, carbon-12, carbon-13, and carbon-14 are isotopes of the element carbon, the number denoting the approximate atomic weights. Isotopes have very nearly the same chemical properties, but often different physical properties (for example, carbon-12 and -13 are stable, carbon-14 is radioactive).

Lethal Dose 50/30 (LD 50/30) The dose of radiation expected to cause death within 30 days to 50% of those exposed. Generally accepted to range from 400 to 450 rem received over a short period of time.

Low level Waste Radioactive waste not classified as high level waste, transuranic waste, spent fuel, or byproduct material. Most are generally short-lived and have low radioactivity.

Medical Physicist A medical physicist has training and experience in the application of radiation to the human body for purposes of diagnosis or therapy. The individual works closely with doctors to make sure that the dose prescribed for a treatment or diagnostic image is delivered. The medical physicist may do a variety of tasks including: perform quality assurance checks; calibrate x-ray equipment; confer with the physician on dosage parameters; check patient charts to make sure that prescribed doses are delivered and to monitor progress; participate in and design radiation protection procedures for staff, patients and visitors; and design room shielding

Microsecond One millionth of a second.

Millirem One thousandth of a rem.

Milliroentgen One thousandth of a roentgen.

Molecule A group of atoms held together by chemical forces. A molecule is the smallest unit of a compound that can exist by itself and retain all of its chemical properties.

Monitoring Periodic or continuous determination of the amount of ionizing radiation or radioactive contamination present in an occupied region, as a safety measure, for the purpose of health protection.

Nanocurie One billionth of a curie.

Neutron An uncharged elementary particle with a mass slightly greater than that of the proton, and found in the nucleus of every atom heavier than hydrogen.

Noble Gas A gaseous chemical element that does not readily enter into chemical combination with other elements. An inert gas.

NORM Naturally Occurring Radioactive Material

NRC (United States) Nuclear Regulatory Commission

Nuclear Energy The energy liberated by a nuclear reaction (fission or fusion) or by radioactive decay.

Nucleus; nuclei (plural) The small, central positively charged region of an atom that carries essentially all the mass. Except for the nucleus of ordinary (light) hydrogen, which has a single proton, all atomic nuclei contain both protons and neutrons. The number of protons determines the total positive charge, or atomic number. This is the same for all the atomic nuclei of a given chemical element. The total number of neutrons and protons is called the mass number.

Nuclide A general term referring to all known isotopes, both stable (279) and unstable (about 5,000), of the chemical elements.

Parent A radionuclide that upon radioactive decay or disintegration yields a specific nuclide (the daughter).

Periodic Table An arrangement of chemical elements in order of increasing atomic number. Elements of similar properties are placed one under the other, yielding groups or families of elements. Within each group, there is a variation of chemical and physical properties, but in general, there is a similarity of chemical behavior within each group.

Personnel Monitoring The use of survey meters to determine the amount of radioactive contamination on an individual, or the use of dosimetry to determine an individual's radiation dose.

Picocurie One trillionth of a curie.

Pig A container (usually lead) used to ship or store radioactive materials. The thick walls protect the person handling the container from radiation. Large containers are commonly called casks.

Planned Special Exposure An infrequent exposure to radiation, separate from and in addition to the annual dose limits.

Pocket Dosimeter A small ionization detection instrument that indicates radiation exposure directly. An auxiliary charging device is usually necessary.

Proton An elementary nuclear particle with a positive electric charge located in the nucleus of an atom.

Rad Acronym for radiation absorbed dose, the basic unit of absorbed dose of radiation. A dose of one rad means the absorption of 100 ergs (a small but measurable amount of energy) per gram of absorbing tissue.

Radiation Area Any area with radiation levels greater than 5 mrem in one hour at 30 cm from the source or from any surface through which the radiation penetrates.

Radiation Detection Instrument A device that detects and records the characteristics of ionizing radiation.

Radiation Shielding Reduction of radiation by interposing a shield of absorbing material between any radioactive source and a person, work area, or radiation sensitive device.

Radiation Sickness (syndrome) The complex of symptoms characterizing the disease known as radiation injury, resulting from excessive exposure of the whole body (or large part) to ionizing radiation. The earliest of these symptoms are nausea, fatigue, vomiting, and diarrhea, which may be followed by loss of hair (epilation), hemorrhage, inflammation of the mouth and throat, and general loss of energy. In severe cases, where the radiation exposure has been relatively large, death may occur within two to four weeks. Those who survive six weeks after the receipt of a single large dose of radiation may generally be expected to recover.

Radiation Source Usually a manmade sealed source of radiation used in teletherapy, radiography, as a power source for batteries, or in various types of industrial gauges. Machines such as accelerators and radioisotope generators and natural radionuclides may be considered sources.

Radiation Standards Exposure standards, permissible concentration, rules for safe handling, regulations for transportation, regulations for industrial control of radiation, and control of radioactive material by legislative means.

Radiation Warning Symbol An officially prescribed symbol (a magenta or black trefoil) on a yellow background that must be displayed where certain quantities of radioactive materials are present or where certain doses of radiation could be received.

Radioactive Exhibiting radioactivity or pertaining to radioactivity.

Radioactive Contamination Deposition of radioactive material in any place where it may harm persons or equipment.

Radioactivity The spontaneous emission of radiation, generally alpha or beta particles, often accompanied by gamma rays, from the nucleus of an unstable isotope.

Radiography The making of a shadow image on photographic film by the action of ionizing radiation.

Radioisotope An unstable isotope of an element that decays or disintegrates spontaneously, emitting radiation. Approximately 5,000 natural and artificial radioisotopes have been identified.

Radiological Survey The evaluation of the radiation hazards accompanying the production, use, or existence of radioactive materials under a specific set of conditions. Such evaluation customarily includes a physical survey of the disposition of materials and equipment, measurements or estimates of the levels of radiation that may be involved, and a sufficient knowledge of processes affecting these materials to predict hazards resulting from expected or possible changes in materials or equipment.

Radiology That branch of medicine dealing with the diagnostic and therapeutic applications of radiant energy, including X rays and radioisotopes.

Radionuclide Radionuclide

Radiosensitivity The relative susceptibility of cells, tissues, organs, organisms, or other substances to the injurious action of radiation.

Radium (Ra) A radioactive metallic element with atomic number 88. As found in nature, the most common isotope has a mass number of 226. It occurs in minute quantities associated with uranium in pitchblende, carnotite, and other minerals.

Radon An invisible, radioactive gas that occurs naturally in soil. Radon is a daughter of radium and one of the heaviest gases known.

Rem The special unit of dose equivalent. The dose equivalent equals the absorbed dose multiplied by the quality factor.

Restricted Area Any area to which access is controlled for the protection of individuals from exposure to radiation and radioactive materials.

Roentgen (R) A unit of exposure to ionizing radiation. It is the amount of gamma or X-rays required to produce ions resulting in a charge of 0.000258 coulombs/kilogram of air under standard conditions. Named after Wilhelm Roentgen, German scientist who discovered X rays in 1895.

Scattered Radiation Radiation that, during its passage through a substance, has been changed in direction. It may also have been modified by a decrease in energy. It is one form of secondary radiation.

Scintillation Detector The combination of phosphor, photomultiplier tube and associated electronic circuits used to count light emissions produced in the phosphor by ionizing radiation.

Secondary Radiation Radiation originating as the result of absorption of other radiation in matter. It may be either electromagnetic or particulate in nature.

Shielding Any material or obstruction that absorbs radiation and thus tends to protect personnel or materials from the effects of ionizing radiation.

Sievert (Sv) The unit of dose equivalent equal to 1 Joule/kilogram. 1 Sv = 100 rem.

Survey A study to:

1) Find the radiation or contamination level of specific objects or locations within an area of interest; and
2) Locate regions of higher-than-average intensity, i.e. hot spots.

Survey Meter Any portable radiation detection instrument especially adapted for inspecting an area to establish the existence and amount of radioactive material present.

SSRCR Suggested States Regulations for Control of Radiation

TRAB Texas Radiation Advisory Board

Terrestrial Radiation A portion of the natural radiation (background) that is emitted by naturally occurring radioactive materials on earth.

Thermoluminescent Dosimeter A device used to measure radiation by measuring the amount of visible light emitted from a crystal in the detector when exposed to radiation.

TRCR Texas Regulations for Control of Radiation

Whole-Body Counter A device used to identify and measure the radioactive material in the body (body burden) of human beings and animals. It uses heavy shielding to keep out background radiation and ultra sensitive radiation detectors and electronic counting equipment.

Whole-Body Exposure An exposure of the body to radiation, in which the entire body, rather than an isolated part, is irradiated. Where a radioisotope is uniformly distributed throughout the body tissues, rather than being concentrated in certain parts, the irradiation can be considered as whole-body exposure.

Wipe Sample A sample made for the purpose of determining the presence of removable radioactive contamination on a surface. It is done by wiping, with slight pressure, a piece of soft filter paper over a representative type of surface area. It is also known as a "swipe sample".

X ray Penetrating electromagnetic radiation (photon) having a wavelength that is much shorter than that of visible light. These rays are usually produced by excitation of the electron field around certain nuclei. In nuclear reactions, it is customary to refer to photons originating in the nucleus as gamma rays, and to those originating in the electron field of the atom as X rays. These rays are sometimes called roentgen rays after their discoverer, W. K. Roentgen.

Appendix 4

Generic Procedures for Building a Radiation Safety Manual

The following procedures can be used by selecting the ones applicable to a given program, editing them to fit the program, and entering the specific information. Question marks (?) are used to indicate the variations that might be encountered. Some programs may require additional procedures. Appropriate regulatory references will need to be researched and entered. This grouping is intended to address the majority of the types of use of radioactive materials.

I. Radiation Protection Program

Note: an adequate operating and safety procedures or radiation safety manual should contain all of the elements required in a "radiation protection program". This method provides an index or summarization. Otherwise, there would be considerable duplication and difficulty in making modifications.

A State Objective: Example—This program is designed to achieve occupational doses and public doses of radiation that are "as low as reasonably achievable" to protect the employees of COMPANY NAME and members of the general public and to comply with [§289.202(e)—or equivalent].

B State Method: Example—COMPANY NAME has established this Radiation Safety Manual (RSM) [or your company's name for the operating and safety procedures] to provide safety guidance to its staff when working with radioactive material. The RSM is composed of? parts:

- Briefly describe your manual's sections and how the sections cover the applicable requirements.

- Indicate the date of implementation
- Indicate when and how the program will be reviewed

C. **Describe the radiation protection program elements**. Include, as a minimum, the following:

- Personnel monitoring requirements and dose limits [ALARA] (reference the sections in your manual that address this area)
- Radiation surveys (references)
- Access controls for radiation areas
- Respiratory protection
- Security of radiation sources (storage/use)
- Posting of areas, rooms, etc.
- Labeling of containers: Example—Sections ? and ? of the RSM provide procedures for labeling of containers (devices and transport containers).
- Receipt of packages containing radioactive material
- Waste storage, processing, transfer and/or disposal procedures
- Management of required records
- Reports of incidents

II—Management Procedures—This section list procedures that address management's control over a radiation use/safety program.

A. **Management Safety Procedures**—This section describes management's radiation safety procedures and controls and supplies most of the information to support the radioactive material license.

1. **Definitions**: (not required, but simplifies the process). Examples—

Agency and BRC:	means the Radiation Control Program, Texas Department of State Health Services [or appropriate agency].
ALARA:	means "as low as reasonably achievable".
DOT or US DOT:	means the United States Department of Transportation.
License:	means the Texas Radioactive Material License issued by the Agency to COMPANY NAME.
RSO and ARSO:	means the ***Radiation*** Safety Officer and the Assistant Radiation Safety Officer, respectively.

2. **General Information: Examples**

- Scope of operations: COMPANY NAME uses (radiation sources/devices) to perform….
- Facilities: COMPANY NAME's main offices and facilities are located in?. All records related to the radiation safety program and the rules are maintained at the main site. For other authorized sites, site specific information will be maintained at the other site(s), where required. A description of the main facility and drawings are found in Attachment?.
- Ownership of property: For each site COMPANY NAME uses to permanently store radioactive material, a document of the property owner's acknowledgment of the presence of radioactive material at the site will be found in the appropriate attachment.
- Management structure and responsibilities—describe duties of owners, managers, RSO, Assistant RSO, consultant, etc. Be sure to discuss the RSO's authority for stopping operations when radiation hazards are encountered.
- Radiation Use Staff and Responsibilities—describe duties and responsibilities of users, trainees, assistants, etc. Indicate supervision of persons not yet fully authorized to use radiation sources.

B. Management Safety Controls and Procedures: this section describes generic procedures and examples for managing the radiation safety program.

1. General safety policies: Examples—

- RSO Authority: The RSO has full authority to stop any operation, involving radiation sources, which jeopardizes the safety of COMPANY NAME employees or members of the general public, or when ordered by the agency, until such time as conditions are made safe. The ARSO has the same authority in the absence of the RSO. The RSO reports directly to the ?.

- RSO (and ARSO) Qualifications—The RSO's qualifications must be, as a minimum:
  - possession of a high school diploma or a certificate of high school equivalency based on the GED test

  - additional requirements regarding training and experience

- RSO (and ARSO) Duties and Responsibilities—The RSO duties and responsibilities include, but are not limited to, the following:

  - establish and oversee operating, emergency, and ALARA procedures, reviewing them regularly to ensure that the procedures are current and conform with the applicable rules;

  - (additional duties and responsibilities)

  - ensure that personnel comply with the requirements of ?, the conditions of the license, and this manual.

  - additional requirements of agency rules and/or reg guides
  - performance of the annual radiation protection program review/audit according to item 20, below.

- ALARA policy: It is COMPANY NAME's policy to maintain radiation exposures to personnel and members of the general public as low as reasonably achievable. This is accomplished through training of personnel to properly handle all radiation producing devices in a safe manner, keeping unnecessary personnel and persons away from areas of use and storage, reviewing personnel monitoring records and counseling employees that may show unusually high exposures, and instilling a sense of responsibility and safety in all staff.

2. Personnel Radiation Monitoring and Protection: Examples—

- Personnel monitoring

    - Use: COMPANY NAME uses personnel monitoring devices to determine radiation worker exposures for normal operations and for emergencies as all employees participating in ? operations are likely to exceed ten (10) percent of the limits specified in ?. Each individual participating in ? (type of operations) or related procedures shall, at all times, wear a film badge or thermoluminescent dosimeter (TLD), according to procedures contained in this manual.

    - Type: Personnel monitoring consists of body badges worn from the belt to the collar (or describe other types). The badges are exchanged according to the prescribed interval (see your section which describes service and intervals) and the personnel monitoring service companys records are audited by the RSO and filed for agency inspection. Each of COMPANY NAME's employees working with radiation shall wear his/her own assigned badge. All control badges for the monitoring system are maintained in a location away from radiation areas under the control of the RSO or a designee. The RSO is responsible for the distribution of the badges and the procedures governing their use. Badges will be returned to the RSO at the end of the wear period for the badge.

- Additional Monitoring Procedures: (be sure to refer to current applicable rules when addressing these areas!)

    - Exposure controls: each employee's exposure history will be obtained to ensure that no worker exceeds any exposure limit.

    - Pregnancy: A woman employee declaring ***her*** pregnancy will have her radiation work duties assessed and her job responsibilities will be modified to prohibit occupational exposures from exceeding 0.5 rem (5 mSv) during the term of the pregnancy.

- Minors: COMPANY NAME does not employ minors for radiation work; therefore no procedures have been developed regarding occupational exposure of minors.

- Planned Special Exposures: COMPANY NAME will not use **planned** special **exposures, pursuant to (cite rule). Therefore no** documentation is required nor maintained.

- Bioassays: COMPANY NAME uses only sealed sources in its operations. Consequently, no internal doses are anticipated and bioassays ARE NOT REQUIRED nor performed.

— or —

COMPANY NAME uses unsealed sources in its operations. Consequently, internal doses are possible and bioassays are used under certain circumstances. A bioassay will be performed when an individual handles more than . . . (describe conditions). The procedure is described in (section ?) and the results are recorded on the (record ?).

- Respiratory Protection: COMPANY NAME does not use uncontained nor airborne radioactive materials, thus respiratory protection is not required nor used.—or describe respiratory protection program.

- Area Monitoring: COMPANY NAME does not routinely monitor areas on a continuous basis. As indicated in section ? of this manual, periodic surveys will be performed around the storage area and surveys will be performed for each ? operation using procedures specified in this manual. COMPANY NAME may use continuous monitoring devices to confirm that radiation levels in unrestricted areas meet regulatory limits when COMPANY NAME may deem it necessary. or describe area monitoring program.

3. Posting of Areas and Rooms: Examples—

- Radioactive Material Use/Storage Areas: a "CAUTION: RADIOACTIVE MATERIAL" sign/label will be placed at the storage facility (bunker) entrance. Any other area where ? containing radioactive materials are used/stored will also be posted—unless an authorized person is continuously present in the area providing surveillance. Signs may also be posted around the perimeter(s) of these areas.

- Radiation Areas: each area, in which radiation levels are such that, if an individual were continuously present in the area, he/she could receive a dose greater than 5 mrem in any one hour (from a radiation source at 30 cm or greater distance), will be posted with a sign that reads "CAUTION—RADIATION AREA". Determination of applicability will be by measuring radiation levels and calculating potential doses or by calculation using the isotope, quantity, gamma factor, and shielding criteria. Signs may also be posted around the perimeter(s) of these areas.

- High Radiation Areas: each area, in which radiation levels are such that if an individual were continuously present in the area he/she could receive a dose greater than 100 mrem (.1 mSv) in any one hour (from a radiation source at 30 cm or greater distance), will be posted with a sign that reads "CAUTION—HIGH RADIATION AREA". Determination of applicability will be by measuring radiation levels and calculating potential doses or calculation using the isotope, quantity, gamma factor, and shielding criteria. High radiation areas should not be developed in COMPANY NAME's operations except momentarily (< 1 hour).

- Operating Areas: Areas where procedures are conducted ? (radiation sources) may be posted with "CAUTION-RADIATION AREA" (CRA) signs according to operating procedures described in ? of this manual. Determination will be according to section ?.

- Review and Inspection of Signs/Posting: the proper posting, and legibility, of caution signs at fixed locations will be reviewed during the annual RPP review and annual internal audit. These reviews may be performed more frequently, but will not necessarily be documented.

- Airborne Radioactive Materials Areas: and so forth . . .

4. **Labeling of Containers and Sources: Examples (useful for transport, esp. well-logging)—**

- Source Holders, Tools, Devices: Each source, source holder, or ? containing radioactive material other than exempt quantity, will be provided with a durable, legible, and clearly visible marking or label that has, as a minimum, the standard radiation caution symbol with no color requirement, and the wording:

DANGER (or CAUTION)

RADIOACTIVE—DO NOT HANDLE
NOTIFY CIVIL AUTHORITIES (OR NAME OF COMPANY)

The labeling will be on the smallest component (i.e., source, source holder, or?) that is transported as a separate piece of equipment, as a minimum.

- Transport Containers: For ? tools/devices, each *associated* transport container:
    - *shall have permanently attached to it a durable, legible, and clearly visible label that has, as a minimum, the standard radiation caution symbol and the wording;*

DANGER (or CAUTION)
RADIOACTIVE
NOTIFY CIVIL AUTHORITIES (OR NAME OF COMPANY)

    - *shall have attached to it a durable, legible, and clearly visible label(s) that has, as a minimum, the licensee's name, address, and telephone number, the radionuclide, its activity, and the assay date and*
    - *shall have two (2) appropriate USDOT labels fixed to opposing sides and the labels shall provide the appropriate information.*

- Label Inspection: each source/tool/device containing radioactive material and its associated transport container will be inspected during the quarterly inspection/maintenance procedures to assure that the labels are properly affixed, legible, and the information is correct. (They will also be inspected prior to leaving for each job.) Sealed ? source/source-holders will be labeled/engraved by the manufacturer and will not be inspected by COMPANY NAME's staff.

5. Certification of Sources: Examples for well-logging—

- Each sealed source that was manufactured after (date in rules) and is used by COMPANY NAME for well logging operations shall have been certified at the time of manufacture to meet the minimum criteria of (regulatory reference). If a certificate is not available and a copy of the prototype testing cannot be obtained, the source will be retested by an authorized service company. The certificate(s)/documentation will be maintained on file.

6. Posting of Notices to Workers: Examples—

- Pursuant to [regulation ?], a notice shall be posted indicating where workers can examine the requirements of [regulation ?]; the radioactive material license, amendments, and conditions and documents incorporated into the license; operating procedures applicable to work under the license; and any notice of violation involving radiological working conditions, or order issued pursuant to [regulation ?], and any response to such notice (see Attachment ?).

- Pursuant to [regulation ?], FORM ?, "Notice to Employees", shall be posted.
- The notices described in ? and ?, above, will be conspicuously posted in close proximity to the storage facility, as a minimum. Additional notices will be conspicuously posted, as necessary and appropriate, in a sufficient number of places to permit individuals engaged in work under the license to observe them on the way to or from particular work locations under the license. The notices will be reviewed for appropriateness, applicability, and proper location during the radiation protection program annual review.

7. Radioactive Material Receipt/Disposal Procedures: Examples—

- Radioactive Material Receipt: (Review applicable rules—may have some variation from agency to agency)

    - Notification of package: Upon receiving notification of the arrival of a package (shipped to COMPANY NAME and containing radioactive material) at a carrier's terminal, COMPANY NAME shall immediately dispatch authorized personnel to pick up the package and properly transport it to COMPANY NAME's authorized facilities.

    - Receipt of package: Upon receiving a package at COMPANY NAME's facilities (whether delivered by common carrier or by COMPANY NAME personnel), the RSO, or his/her trained designee, shall examine the package within three (3) hours of receipt for proper labeling of the package and for damage or loss of structural integrity to the package or its contents. Packages received after normal working hours shall be stored in the storage facility and examined within three (3) hours from the beginning of the next work day or shift.

- Package monitoring: the exterior of the transport container shall be surveyed, and the results recorded on (record ?), to assure that radiation levels do not exceed 200 mr/hr; on the surface of the container.
- Package discrepancies: COMPANY NAME shall notify the Agency if any discrepancies of the package labeling are found. COMPANY NAME shall IMMEDIATELY secure the package in the storage area and notify the Agency and the carrier if the package and/or its contents (describe the anticipated radioactive material or devices) appear to be damaged.
- Received radioactive material use: received (devices) containing radioactive materials WILL NOT be put into, or returned to, service unless accompanied by a certificate from the shipper/transferor indicating that the sealed sources contained within were tested for leakage within the preceding six months. In such a case, COMPANY NAME will perform a leak test, using procedures described in (section ?) of this manual. If the test results show that the radioactive source(s) are not leaking, then the (device) containing radioactive material will be placed into service.
- Receipt record: A receipt record (see ? record) will be completed and filed for each received device containing radioactive material.

- Disposal—Radioactive material will be disposed of by shipping it to the authorized supplier or manufacturer. As an alternative, a licensed waste disposal service company may be used (see section ?) as a waste disposal resource. The (record ?) will be used as a record of disposal.

8. **Inventory: Example**—COMPANY NAME performs and records [quarterly, 6 months, or annual?] physical inventories. The inventory record is shown in (section or attachment ?). The inventory process consists of locating each specific radiation source (tool/device) and documenting that it remains in COMPANY NAME's physical possession and control.

9. **Waste Disposal: Examples—**

- Waste Disposal: COMPANY NAME does not generate radioactive waste in its operations. Radioactive material will be returned to the licensed manufacturer when no longer serviceable or when replaced by newer devices or sources. In the event that waste is generated (due to an accident, for example) or the radioactive material cannot be returned to a licensed manufacturer, COMPANY

NAME WILL use the services of a licensed radioactive waste broker/disposal service company for proper disposal. The proposed service company is identified in (section ?). (Note: COMPANY NAME reserves the right to change service companies if circumstances warrant.)

— or —

- Waste Disposal: Radioactive waste will be disposed of by decaying and discarding it in normal waste streams; by returning it to the authorized supplier or manufacturer; or, as an alternative, a licensed waste disposal service company may be used (see section ? or attachment ?) as a waste disposal resource. The (record ?) will be used as a record of disposal. The procedures are described in (section ?).
  - Example Waste Disposal Procedure (for short lived isotopes): Two methods of disposal will be used—
    - Decay Procedure—COMPANY NAME will dispose of radioactive wastes using the following procedure:
      - Seal the container, place it in a heavy plastic bag, and seal the bag tightly.
      - Clearly mark the closure date on the bag.
      - Determine the decay date by adding (time period based on 7-10 half-lives) to the closure date and clearly mark it on the bag.
      - Once each month, remove the bags for which the storage for decay time has expired (as indicated by the decay date on the bag), remove or obliterate any radiation warning labels, survey the package (away from the storage area), and, if the survey shows that decay is completed, dispose of the package in normal waste. The package should not be placed in waste streams destined for incineration.
    - Disposal: Alternative to the above, radioactive waste may be disposed of by returning it to the authorized supplier or manufacturer. As a further alternative, a licensed waste disposal service company may be used (see section ?) as a waste disposal resource.

- Record: the (record ?) will be used to document waste disposals.

  Note: Texas has a "300 day half-life" rule which allows discarding radioactive wastes below certain activities/concentrations. The authorizations afforded by other regulatory agencies should be carefully reviewed before incorporating a procedure like the one above.

10. Storage of Radioactive Material: Examples—

Note: If there is more than one authorized site for storage, be sure to address each site individually!

- Storage Facility: The storage facility consists of a storage bunker located in (describe). The bunker is a (describe construction). The surrounding/adjacent areas are of (?) occupancy and there are (?) "full-time" work stations of unmonitored personnel near the storage facility, the closest station on site being at least (? distance) away. There is (?) combustible material in the storage area so the potential for damage to (radiation sources) by fire is extremely (?).

  Note: if there are other hazards in the area, such as explosives, be sure to address the methods of assuring overall safe storage.

- Storage Security: There are at least (?) "levels of security" for each stored device—container lock, storage bunker lock, and fence. When not in use and not installed securely in a (vehicle), all tools/devices containing radioactive materials are stored within the designated storage facility with (describe storage and locking mechanism). The storage (?) is kept locked when personnel are not present storing or removing radiation sources. Keys to the storage facility are maintained by the RSO. The RSO shall control the issuance of storage facility keys to assure complete security at all times. Only authorized users may check out a key to gain access to radiation sources.

11. Control of Exposure from External Sources and Other Engineering Controls: Examples—

- Restricted Areas: the areas (see drawings in section ?) restricted for purposes of radiation protection are, as a minimum:

- Storage facility: the storage facility and the area . . .—the (example—fence) and company personnel control and/or prohibit members of the general public from entering any area where radiation is, or may be, present.

- Use areas: the areas (be sure to address temporary job sites) where (radiation work?) is performed—during ? operations, company personnel visual controls are used to prohibit unauthorized persons from entering the area(s). On occasion, warning signs may be used.

- High Radiation Area Controls: accessible areas in which radiation levels could result in an individual receiving a dose equivalent greater than 0.1 rem (1 millisievert) in one hour at 30 cm from any source of radiation or from any surface that the radiation penetrates exist (describe conditions). Direct surveillance by the (radiographer, well-logging supervisor, technician, etc.) will be used to prohibit persons from entering the area(s). Be sure to review regulatory requirements for controls (interlocks, etc.) applicable to the particular type of operation.

- Other Engineering Controls: COMPANY NAME does not use unsealed radiation sources. No other engineering controls are required or used.

— or —

- describe the controls to be used (air handling systems, fume hoods, etc.)

12. Radiation Surveys: Examples—

- Radiation surveys: will be performed, using a currently calibrated radiation survey instrument appropriate for the type of radiation, and the results will be recorded on forms shown in (section ?) of this manual. As a minimum, the following surveys will be routinely performed at the indicated frequency or time(s):

- Storage surveys: a radiation survey of the storage facility will be performed [? time period, such as quarterly] (consider: usually at the time of inventory insertion here) and the results will be recorded at the time of the survey. Should one or more of the radiation sources/tool/devices not be available due to their use in locations distant from the storage facility, the survey will still be performed and be considered as representative of the general storage conditions—although every effort will be made to have all sources stored at the time of the survey. Should an additional tool/device containing radioactive material be added to the inventory, a storage facility radiation survey will be performed and recorded. The surveys will be performed by the RSO, ARSO, or a qualified (user) as assigned by the RSO.

- Radiation levels in unrestricted areas—Should a radiation survey determine radiation levels in an unrestricted area exceeds regulatory limits:
    - the RSO will promptly restrict the area(s) until storage conditions can be changed to reduce the levels to within limits; OR
    - the RSO may use occupancy factors to show by calculation that no member of the general public will be exposed above the limits and then change the restricted area boundaries and/or shielding properties in a timely manner.
- Determination that limits may be exceeded: A continuous radiation level of 1.8 mr/hr or more at a location in an unrestricted area shall indicate that the limit of 2 mrem (0.02 millisievert) in any one hour may be exceeded and that the condition deserves evaluation to determine an appropriate action.

- Use Area Surveys: are generally performed (describe conditions). The (record ?) is used to record the results.

- Tool/Device Surveys: are performed when securing the (tool/device) from use if the radioactive source is removed from the tool to assure that there is no contamination present and a survey would also be performed if the tool/device appears to be damaged.The results are recorded on (record ?).

- Vehicle Surveys: vehicle surveys will be performed when transporting radioactive material or tools/devices containing radioactive materials and recorded according to transport procedures specified in this manual [see section?]. The (radiographer, well-logging supervisor, technician, etc.) transporting the tool/device will perform and record the survey.
    - (in Texas): Vehicles loaded with containers of radioactive material will not be parked in "residential locations" unless this survey has been properly performed and the results recorded.

- Public dose estimates will be established for each storage and/or use area and the results will be documented.

- Records of Radiation Surveys:

  - Storage Areas: The results of storage facility surveys will be recorded on the (record ?). Note: be sure to address storage facilities at other sites.

  - Use area surveys: When a use area survey is performed, the results will be recorded on (record ?).

  - Vehicle surveys: will be recorded on (record ?), as appropriate.

13. Radiation Survey Instruments: Examples—

- Radiation survey instruments: are available to perform radiation surveys during radiation use operations and for surveys of the storage facilities and responses to emergencies.

  - Types of Instruments: COMPANY NAME possesses a sufficient number of GM instruments to have at least one calibrated instrument of the appropriate type available to be used for storage surveys, vehicle surveys, and during radiation use procedures. See section ?

  - Instrument Specifications: Each instrument used in (?) operations will be capable of measuring from (?) milliroentgen per hour through (?) milliroentgens per hour, as a minimum.

  - Instrument Calibration: Each radiation survey instrument shall be calibrated at intervals that do not exceed six (6) months [or specify other time period] and following each instrument servicing or repair other than battery replacement. COMPANY NAME will require that each instrument be calibrated to an accuracy of 20 percent of the true radiation level by a service company authorized to perform instrument calibration. The service company used by COMPANY NAME is identified in section ?.

  - Instrument Use: Radiation survey instruments that have exceeded the calibration interval <u>will not be used</u> until they have been successfully recalibrated.

14. Records Management: Examples—

- Main site: the RSO shall collect, file, and maintain all license, registration, and radiation safety related records at the main facility, for at least the minimum time(s) specified by (applicable rules). The records shall be made available for agency inspections and for emergencies. These records include:
    - Use/Storage Log: describe record, when and how it is use, where it will be filed, where a copy of it is found (in the manual), and how long the record will be maintained.
    - Receipt and transfer records
    - Leak test records: Example—consist of the service company certificates that are maintained in a specific file for each source.
    - Personnel monitoring reports
    - Instrument calibration records
    - Radiation survey records
    - Training records
    - Inventory, inspection, and maintenance
    - Radiation protection program annual review
    - Personnel annual evaluation record
        - Other records listed in procedures
- Other Authorized Sites: for other authorized sites, the records that will be maintained will be identified in section ?
- Temporary Job Sites: the following records, or copies, will be available and maintained at each temporary job site:
    - a copy(s) of the radioactive material license.

- a copy of Parts III? & IV? of COMPANY NAME's Radiation Safety Manual (which includes operating and emergency procedures).
- a copy of (list applicable rules)
- Radiation survey records for the period of operation at the job site, should any be performed.
- The current instrument calibration and leak test records for instruments and sources, respectively, being used at the job site.
- Records of transport (shipping papers).

15. Transport of Radioactive Material: Examples—

- Transport Modes: Tools/devices containing radioactive materials are transported in COMPANY NAME vehicles in transport containers stored in secure compartments on the vehicle. The container is secured with a lock to prohibit unauthorized removal and blocking and bracing is provided, when necessary, to prevent movement within the vehicle during transport.

- Transport Procedures: Radioactive material transported in COMPANY NAME's own vehicles will have the tool/device locked, will be placed in an appropriate, properly labeled US DOT transport container, and the transport container will be locked, blocked, and braced—as appropriate—on the vehicle. Shipping papers and vehicle placarding will be provided, as follows:

  - Transport Containers and Labels: Only containers which meet US DOT requirements and which are properly labeled according to US DOT will be transported. Two (2) [White I, Yellow II or Yellow III] labels will be affixed to opposite sides of each transport container. The labels will each indicate: the isotope, quantity, and transport index.

  - Vehicle Placarding: Vehicles will be placarded with the US DOT "diamond shaped" signs specifying RADIOACTIVE during transport—if the container is required to be labeled with "Yellow III" labels. A placard will be placed on each of the four (4) sides of the vehicle in a position that allows them to be clearly visible.

- Shipping Papers: A record (shipping paper) identifying the radioactive material, source activity, transport index, etc., will be carried at the driver's position and in immediate reach of the driver during transport. Shipping papers (bill of lading) and placards will be provided to the drivers of vehicles of commercial carriers, when that mode of transport is used. See record (?)—shipping paper.

- Vehicle Surveys: Radiation surveys of vehicles will be performed and recorded, when required, according to procedures provided in this manual. ? will be used for the record.

- Records of Transport: Each transfer/transport of radioactive material will be documented on COMPANY NAME's "Radiation Source Utilization Record" or "Receipt/Transfer Record", as appropriate and/or applicable.

- Transport Index Correction: the sources used by COMPANY NAME have relatively long half-lives so the transport index will not require appreciable changes over long time periods.

16. Equipment Inspection and Maintenance Procedures: Examples—

- Equipment Inspection: The specific procedures and documentation are found in the indicated attachments for inspection of the following:

  - Six Month (or other interval) Equipment Inspection: At intervals no greater than six (6) months (or other interval), the RSO will direct inspection, review, maintenance and documentation of (devices) containing radioactive materials and related equipment. As a minimum, the procedure shall include review/inspection of:

    - proper labeling of the tool/device and its associated transport container;

    - proper condition and labeling of the source and its container; and

    - proper operation/condition of associated handling tools.

  - Maintenance: minor maintenance (such as cleaning exterior surfaces) is performed on tools/devices containing radioactive materials by authorized (radiographer, well-logging supervisor, technician, etc.) as determined by the periodic inspections. Major maintenance/repair (requiring removal of source) is only performed by the manufacturer or a party specifically licensed to perform such procedures.

- Defective components: To assure safe operation, defective, inoperable, or otherwise unserviceable equipment and associated components described above will be removed from service and tagged as unusable until they have been repaired.
- Inspection and maintenance records: the inspection, maintenance, and subsequent repair shall be documented on record ?.

17. Emergency Procedures: Examples—(general descriptions of procedures—expanded procedures can be described in more detail in a specific section—these may be sufficient for many programs)

- Emergency procedures: for radiation emergencies are provided in Section ? of this manual. The procedures address responsibilities of both management and the user.

(Choose possible types of emergencies)

- Types of Emergencies: The following procedures are addressed:
    - General field emergency procedure.
    - Theft of a source/tool/device.
    - Loss of a source/tool/device.
    - Emergency due to source/tool/device/equipment malfunction.
    - Emergency due to vehicle accident.
    - Emergency causing exposure of unmonitored persons.
    - Emergency involving fire.
    - Emergency source recovery of a "stuck tool" (well logging).
    - Emergency due to leaking source

- Emergency due to contamination of persons, facilities, and/or equipment

(Example procedures are listed in Chapter 4 and at the end of this appendix.)

- RSO Responsibility: For the above, and any other radiation emergency, the RSO is responsible for timely contacting all appropriate authorities, assuring the site of the emergency and the tool/device are secured and persons are protected from radiation hazards, performing necessary leak tests, and arranging for proper shipment or disposal of the tools containing radioactive material or device, if necessary. The RSO is also responsible for documenting the incident and providing the appropriate reports to the Agency. The services of a qualified consultant in radiation safety may also be used.

18. Leak Test Procedures: Examples—

- Authorized person: The RSO, ARSO, radiation safety consultant, or a qualified (radiographer, well-logging supervisor, technician, etc.), will be the only persons authorized to perform wipes for leak tests of sealed sources. An agency authorized/licensed leak test service company/person/consultant may also perform the tests.

  Schedule: Leak tests will be performed on sealed sources at intervals no greater than six (6) months (check applicable rules for variations in time intervals) using kits from a service company authorized by the agency. They may be performed more frequently if circumstances warrant.

- Service company: The service company used by COMPANY NAME to provide for leak tests and analyses is identified in Attachment ?.

  (Note: COMPANY NAME reserves the right to change service companies if circumstances warrant.)

- Procedure: The steps for performing a leak test are:

  - the date of the test, source serial number, isotope type, and isotope quantity are recorded on the service company's leak test form/record;

  - using the specific instructions provided by the leak test survey company with its leak test kit, one or more wipes (swabs) are obtained and placed in the kit's plastic bag(s) or container(s);

- the source is secured; and (note: it may be necessary to expose the source in some cases)
- the wipes/swabs are removed from the storage area, surveyed with a calibrated radiation survey instrument, and the following considered:
  - if the radiation survey does not indicate leakage, the swab/wipe is returned by mail, commercial carrier, or COMPANY NAME personnel to the service company for analysis; or
  - if the radiation survey does indicate a positive result (i.e. potential leakage of the source), the RSO will immediately withdraw the source/ tool containing radioactive material from service, secure it in the storage area, and notify the Agency, the manufacturer's representative, and the consultant. The contaminated leak test swab will be sealed and stored in the storage area pending determination of the proper course of action. The swab/wipe WILL NOT be presented for mailing or for shipment by common carrier.

- Leak Test Results: the RSO will review the leak test certificate/record returned from the service company:
  - if the leak test results are negative (i.e., no leakage), the record will be filed for inspection; or
  - if the leak test results are positive (i.e., an indication that the source may be leaking), the RSO will immediately withdraw the tools/ devices containing radioactive material or device from service, secure it in the storage facility, and notify the Agency, the manufacturer's representative, and the consultant. The device will be kept in the storage facility pending determination of the proper course of action. A written report will be provided to the Agency.

19. Training and Safety:

- Training Program:
  - Employees will not participate in (radiation use) activities without supervision by a qualified (radiographer, well-logging supervisor, technician, etc.) until they have:

    - received, and successfully completed, a ? hour (minimum) training course, conducted by an authorized training service company or consultant, which includes the subjects outlined in (applicable rule or reg guide);

    - received and reviewed a copy of the rules contained in (applicable rules) and . . . ;

    - received and reviewed a copy of the license and conditions ;

    - received and reviewed a copy of the operating and emergency procedures;

    - demonstrated understanding of the regulatory requirements and the requirements of this manual by successfully completing a written examination given by COMPANY NAME;

    - completed ? hours/days/weeks/months of on-the-job training under the supervision of a qualified (radiographer, well-logging supervisor, technician, etc.);

    - *demonstrated through a field evaluation, given by COMPANY NAME, competence in the use of sources of radiation, related handling tools, and the type of radiation survey instruments that will be used in the job assignment;.and*

    - *a (training record ?) form has been completed and filed.*

  - Employees will not participate in ? activities as an ? assistant until they have:

    - completed the steps of (sections ?) of this manual; and

    - demonstrated competence to use, under the personal supervision of ?, the sources of radiation, related handling tools, and radiation survey instruments that will be used in the job assignment.

- Training Records: The RSO will maintain records showing the training, qualification, and proficiency of each radiation user.

- Prohibitions:

  - No employee or person shall operate or work with any of COMPANY NAME's ? tools/devices containing radioactive materials until that employee or person has completed the required training and has received the approval and authorization of the RSO.

  - Any operation, such as drilling, cutting, or chiseling on a source holder containing a sealed source, WILL NOT be performed by COMPANY NAME employees. Nor shall they perform the repair, opening, or modification of any sealed source.

20. Program and Personnel Reviews:

- Program Internal Audit: an annual radiation protection program review will be performed during the anniversary month of the inception of the radiation protection program and each year thereafter. The reviews will be performed by the RSO with assistance from the ARSO and experienced, qualified staff, and/or by the consultant. The review will assure that the radiation protection program is in compliance with current regulations and that personnel are performing their duties in compliance with the regulations and license requirements. The results are recorded on record ?.

- ? Personnel Annual Evaluation: a review of each person (radiographer, well-logging supervisor, technician, etc., and assistants) participating in ? operations shall be performed annually. The review of each individual's performance will be conducted by the RSO or the ARSO, and/or by the consultant. The items reviewed are shown on record ?.

- Deficiencies: deficiencies found during any audit or evaluation, and corrective actions taken, will be recorded on record ? and reported to the (president, company official) of COMPANY NAME.

III—User/Operator/Worker Procedures—*This section list procedures that address workers activities during operations under a radiation use/safety program. It can be used to prepare a "manual" for workers to have available during operations. The manufacturer's procedures for the specific equipment being used can be added.*

A. General Safety Requirements for Radiation Workers

1. Employee/Radiation Worker Responsibilities and Duties: Examples—

- Prohibition: No employee or person shall operate or work with any of COMPANY NAME's devices or tools containing radioactive materials, until that employee or person has completed the required training program, for the particular level of work, and has received the approval of the Radiation Safety Officer (RSO). Procedures requiring specific authorization are:
    - Performance of ? using ? sources, tools, and/or devices;
    - Performance of leak tests; and
    - Performance of repairs or maintenance to radiation sources and devices.
- Prohibition: Any operation, such as drilling, cutting, or chiseling on a source holder containing a sealed source, shall ONLY be performed only by persons specifically licensed to do so by the agency, another agreement or licensing state, or the NRC. Further, the repair, opening, or modification of any sealed source shall be performed ONLY by persons specifically licensed to do so by the agency, another agreement or licensing state, or the NRC. COMPANY NAME employees WILL NOT perform any of these functions!

2. ALARA: it is COMPANY NAME's policy to maintain radiation exposure to "as low as reasonably achievable". Each of COMPANY NAME's employees will take every available precaution to protect himself/herself, fellow workers, and members of the general public from unnecessary radiation exposure while working with or operating tools/devices containing radioactive materials.

3. Employee Responsibility: Each COMPANY NAME employee (radiation worker) will keep this document (applicable parts of COMPANY NAME's Radiation Safety Manual) available for reference, instruction, and guidance while working with radiation sources and will abide by ALL requirements of the document.

4. Failure to Follow Safety Procedures: any employee who threatens the health and safety of fellow employees or the general public by failing, whether negligently or willfully, to follow or comply with the following will be subject to disciplinary action by the company:

- all safety requirements presented by this manual;
- oral or written instruction or directives from COMPANY NAME'S management;
- rules and requirements of the applicable (rules/law); and
- oral or written directives from officials of regulatory agencies with jurisdiction.

5. Worker responsibilities for ? operations:

- Duties, prohibitions, and restrictions of (radiographer, well-logging supervisor, technician, etc.):
    - ? supervisors are responsible for any radiation sources/devices assigned to them;
    - ? supervisors are responsible for understanding the policies of this company and of the regulatory agencies;
    - ? supervisors are responsible for having all required safety equipment on hand prior to beginning operations and throughout all procedures;
    - ? supervisors are responsible for complying with all procedures, on a daily basis, when performing radiation use operations and related safety procedures; and
    - ? supervisors will provide responsible supervision of assistants, when assigned.

    Note: The ? represents the fully trained and qualified individual for that type of use (radiographer, radiographer trainer, well-logging supervisor, etc.)

- Duties, prohibitions, and restrictions for (? assistants)
    - ? assistants shall only perform ? operations as directed by, and under the supervision of, a ? supervisor;
    - ? assistants are responsible for understanding the policies of this company and of the regulatory agencies;

- ? assistants are responsible for having all required safety equipment on hand prior to beginning ? operations and throughout all w? procedures; and

- ? assistants are responsible for complying with all procedures, on a daily basis, when performing ? operations and related safety procedures.

B. Personnel Safety Requirements for Radiation Workers: Examples—

1. Personnel Monitoring Requirements: Personnel monitoring will be used as designated in this manual. No employee or person will operate any radiation source, nor be involved in such operation, unless he/she is properly equipped with the personnel monitoring equipment required by this manual.

- Wearing of Badges: Each radiation worker will wear his/her assigned film badge or thermoluminescent dosimeters (TLD), and ONLY his/her own badge, at all times during procedures or activities involving radiation sources. The badge shall be worn in a position from the waist to the collar and towards the front of the body.

- Care and Control of Badges: Each radiation worker will remove his/her assigned film badge or TLD at the end of the work period and store it in a location designated by the RSO—away from areas of potential radiation exposure to the badge. Each worker shall take particular care to protect badges from damage by heat, chemicals, inadvertent radiation exposure when not worn, etc.

2. Worker Safety Equipment: Each radiation worker will use the following equipment according to procedures specified in this manual:

- Radiation Survey Instruments: During ? operations and operations involving radiation sources (such as storage surveys), each radiation worker will use a calibrated radiation survey instrument provided by COMPANY NAME. Radiation survey instruments that have exceeded the calibration interval will not be used until they have been successfully recalibrated. In general, GM instruments are used for any operation with radioactive material.

    - Instrument Operation: Prior to beginning operations or transporting sources to a job site, the instrument shall be inspected and checked for proper operating condition.

- Radiation Survey Instrument Operating Condition Inspection Procedure:

Note that can be included in procedure: As a reminder, exposure rates are determined by multiplying the number that the meter needle points to by the selector switch range position number.

  - Check for Current Calibration: Check the sticker on the side of the instrument to ensure that the instrument is in current calibration. IF THE CALIBRATION HAS EXPIRED, THE INSTRUMENT MAY NOT BE USED!

  - Check the batteries. If the instrument has a battery check position on the range selector switch, turn the selector switch to this position and check the meter to see that the needle indicates that the batteries are good. If the instrument has a test button, turn the selector switch to the appropriate selection, if any, and press the test button while observing the meter needle for an indication that the batteries are good. If the batteries are not good, THEY MUST BE REPLACED!

  - Check Instrument Operation: Turn the instrument selector switch to the "X1" position and hold the instrument (or its detector) next to a source of radiation, such as a check source) to determine if it detects radiation. If the instrument does not detect radiation, IT MUST NOT BE USED! Label the instrument as unusable and turn it in to the RSO.

Note that can be added: the instrument operator should use his/her experience to assess the response of the instrument to the expected radiation level.

  - Instrument Failure: If the survey instrument fails to pass ANY of the inspection check points above, it must be tagged, labeled, or marked "Out of Service" and turned in to the RSO. It MUST NOT be used in any operation!

- Radiation Survey Instrument General Use Requirements—ALL personnel must use radiation survey instruments as follows:

  - Generally, ? operations will not be performed without having available at least one readily available survey instrument in proper operating condition and in current calibration.

- Radiation survey instruments shall be used when and where these procedures direct.

- Personal Safety: Employees will use hard hats, steel toed shoes and other general safety equipment, if provided/required by COMPANY NAME or by the client.

C. Storage Security: Examples—

1. Employee responsibility: Each employee (radiation source user) is responsible to assure that sources under their control are safely stored and secured in the appropriate storage facility when the work or activity is completed.

2. **Radioactive Material Storage Procedure:**

- Store radioactive material in the appropriate container or transport container while not in use.

- Lock the source container and remove the key.

- Inspect the container exterior and assure that it is undamaged and that all required labels are properly affixed.

- Place the container in the assigned storage position within the storage facility and perform a radiation survey to assure that radiation levels are "normal".

- Close and lock the storage facility.

- Provide the required information on record ?.

- Notify the RSO of any observed damage or missing/illegible labeling.
- Notify the RSO of any problems observed with the tools/devices, transport containers, associated operating and safety equipment, or the storage facility.

3. **Procedure for Storage at Temporary Job Sites:** If the radioactive material is to be stored on the vehicle overnight, perform the following:

- Inspect the transport container and assure that it is undamaged and that all required labels are properly affixed.

- Inspect the tool/device exterior and assure that it is undamaged and that all required labels are properly affixed.

- Place the locked tool/device in the transport container, if applicable, and secure the container.

- Place the transport container at the assigned position in the bed or compartment of the vehicle.

- Secure the transport container to the vehicle with chain and lock or provided security equipment.

- When transporting tools/devices containing radioactive materials, perform and record a vehicle survey. Record the results on record ?.

  Note that can be added: Vehicles loaded with containers of radioactive material may not be parked in "residential locations" unless this survey has been properly performed and the results recorded.

- Ensure that the vehicle placards are displayed, if required.

D. Transport Procedures: Examples—

Note that can be added: Radioactive material shall only be transported in COMPANY NAME's vehicles or by commercial carriers.

1. Loading the vehicle: before proceeding to the job (all radioactive material radiation sources):

Note that can be added: Wear personnel monitoring!

- Upon removing the tool/device from the storage facility, inspect the transport container (or overpack) and assure that it is undamaged and that all required labels (2, each on the opposite side from the other) are properly affixed.

- Survey the exterior to assure proper radiation levels and record the results on the record ?.

- Place the transport container at the assigned position in the bed of the vehicle or storage compartment and, when applicable, and lock the compartment.

- When applicable, secure the transport container to the vehicle with chain and lock or other company provided security device(s) to prevent tampering and unauthorized removal;

- If not using a special storage compartment, place blocking/bracing devices to prevent movement of the container during transport.

- If the package (transport container/overpack) has Yellow III labels, display the diamond shaped placards—one on each of the four sides of the vehicle. Assure that the signs are legible and unobstructed from view in each of the four directions.

- Perform and record a vehicle survey and document on record ?.

- Ensure that the correct shipping paper is available and place it in the vehicle cab within immediate reach of the driver.

- If transferring the device to a commercial carrier, ensure that the driver of the vehicle is provided with a shipping paper and emergency phone number(s) and procedures; that he/she properly blocks and braces the package; and, if a "Yellow III" label is required, placards the vehicle with 4 "RADIOACTIVE" diamond shaped placards—one on each of the four (4) sides of the vehicle.

- Complete the proper record (Utilization Log or Receipt/Transfer Record) prior to transport.

2. **Vehicle Survey Procedure**—After securing the device in its transport container or storage compartment on the vehicle, perform the following:

   - Using a radiation survey instrument in current calibration, survey the vehicle on each side at 3 feet from the surface;

   - Survey the driver's position, and, the positions of any passengers;

   - Record the survey results on record ?.

3. Transport of tools/devices containing radioactive materials:

Note that can be added: Drive carefully !!!

- Periodically, check the blocking, bracing, and security of the transport container, as well as the placards and your personnel monitoring.
- When stopping at a commercial facility (such as a restaurant), select a parking place away from the activities of other persons—preferably a location which is reasonably secure and lighted and where visual surveillance can be readily maintained.
- Upon arrival at the destination, immediately check the security and condition of the transport container.

4. Loading the vehicle at the job site: the same basic steps as above.

E. Safe Operation Procedures: Examples—

Note: Procedures during actual operations are highly specific to the type of use. Only a few general steps will be listed here.

Note that can be added: The following equipment must be present prior to making any exposure:

* *Film badge or TLD*
* *Survey instrument*
* *Radiation warning signs*

DO NOT initiate or conduct ? operations without this equipment available and fully operational!

1. General Operational Safety Requirements:

- Preparation for Safe Operation—Equipment Inspection: Prior to removing radiation sources from storage, transport of sources to a temporary job site, or initiating any ? operations, the following steps will be taken:
    - Inspect your personal protective equipment:

- Check film badges for visible damage, assure that the film in badge the holder is current, and check for proper wearing position on the body; and

- Check the radiation survey instrument(s) to assure that the there is no visible damage, the calibration is current (less than ? months since last calibration), the batteries are in good condition, and the instrument responds to radiation sources.

- Perform the equipment inspection procedure for sources/tools/devices, as follows:

  - Check that radiation levels are "normal" when the tool/device is removed from the storage facility;

  - Check the tool/device and container good condition and for proper labeling,

  - Check the general condition of source, tool, and container.

    Note that can be added: Any deficiencies found during the equipment inspection shall be brought to the attention of the Radiation Safety Officer immediately, so that they may be corrected before the equipment is used. Each defective component shall be labeled to provide warning that the equipment is unusable.

- Establishing Use Area Controls:

  - General Security and Safety—The following procedures will be used to ensure maximum security of the exposure area at all times:

    Note that can be added: Never leave a source of radiation unattended!

    - The ? supervisor shall obtain the key to the container when removing it from storage and retain the key in his possession at all times until the unit is returned to storage or transferred to another ? supervisor.

    - After set up, during use, and between uses, the ? supervisor (and/or assistant) will maintain continuous direct surveillance of the area of operation to protect against unauthorized entry into the exposure area.

- Restricting Area Access—maintain constant surveillance of the area, while the source is outside of the transport container and has not been placed down-hole, to prohibit access by unauthorized persons. If necessary, display Caution—Radiation Area signs to warn unauthorized persons.

- Use of barriers, ropes, tapes, interlocks, etc. can be addressed here.

- Radiation Surveys:

  - Radiation Surveys Upon Completing Job: A radiation survey shall be performed at the completion of the job if there is any indication that the source may have been damaged. A record (is or is not) required.

  - Vehicle Surveys at End of Job:

    - after completing the job, store and secure the tool/device in the transport position of the vehicle according to section ?.

    - perform a radiation survey of the vehicle according to section ?.

    - record the survey results on record ?.

- Records of Operations:

  - Utilization Log or Radiation Source Utilization Record, must be completed for each use of radiation producing equipment. The information required by this form is as follows:

    - Removing a Radiation Source from Storage:

      - Enter the date removed from storage;

      - And so on . . .

    - Returning a Radiation Source to Storage:

2. **General Procedure for Performing ? Operations with Tools Containing Radioactive Material:** (Select applicable and add others necessary to your specific operations)

- All sources will be kept in an approved source storage assembly (or transport container) at any time they are riot in use.
- Radioactive materials will not be handled except by personnel authorized by the license and according to these procedures.
- Clear the area of unauthorized/unmonitored persons while transferring the tool/device from the transport storage container to the use location and maintain constant visual surveillance until the transfer is complete. Assure that all personnel within the restricted area wear current personnel monitoring devices at all times.
- Use only the special handling tool(s) to maintain a distance of at least ? feet from the source. Radioactive material/sources WILL NOT be touched or handled by "hand".
- Perform the transfer procedure quickly but carefully! Minimize personnel exposure.
- Inspect any cabling and connectors to assure proper condition. Make sure connections are complete and secure.
- When operations are completed, quickly, but carefully, return the source to the storage/transport container. Minimize personnel exposure.

F. Records: (Records for which the user/operator is responsible, as applicable)

- User Records: enter all of the required information on the Radiation Source Utilization Record form when removing a tool/device from storage and/or when returning it to storage and all of the required information on ? Job Record, before, during, and at the conclusion of, field operations.
- Record Content: enter all of the required information on test records when conducting operations using tools/devices containing radioactive materials.
- Record Accuracy: Each ? supervisor shall review his/her records for accuracy and turn them in upon return to the office/storage facility.
- Other entries . . .

G. **General Emergency Procedures:** *the following example can be used for uses with minimal requirements (a large program may want to have an entire section devoted to emergency procedures)—*

- Secure area: The operator will secure the area of the emergency by establishing a visible barrier at a minimum radius of ? feet from the center point of the tool/device location;

- Contact RSO: The operator will immediately contact the RSO by two-way radio or by telephone and report the circumstances of the emergency;
- Inspect damage: If instructed by the RSO, the operator will inspect the tool/device for damage, otherwise, the RSO will travel to the site and inspect the tool/device;
- Minor damage: If the damage appears to be minor, upon AUTHORIZATION by the RSO, the tool/device may be placed in its transport container and returned to the storage area for an evaluation and emergency leak test;
- Serious damage: If the tool/device appears to be severely damaged, i.e. broken, crushed or burned, a minimum of a ? foot barrier around the tool/device will be established and all persons will be barred from the area. The operator will remain physically present at all times to prohibit entry into the restricted area by unauthorized/unmonitored persons.
- Records: If the RSO is not immediately present at the site of the accident, the operator shall record all of his/her observations regarding the conditions of the emergency, including, but not limited to, the following:

    a. The date/time and conditions at the time of the accident; and

    b. The names, addresses, and phone numbers of persons present (within ? feet of the location of the device involved).

Note: Additional emergency procedures are discussed in Section 3 of Chapter 4 and in the next section of this appendix.

IV—Examples of Emergency Procedures:

A. General Field Emergency Procedure—

User Responsibility/Procedures

- If possible, the radiation source must immediately be placed in its safe position, or, in the case of x-ray machines, turned off and/or the power removed.

- The user should immediately clear all unauthorized/unnecessary persons from the area and then secure the area of the emergency by establishing a visible barrier (such as using warning tape, rope, signs) around the point of the accident (location of the radioactive material—or x-ray machine if power cannot be terminated).

- The user should immediately contact the RSO by telephone or radio or for guidance and instruction.

- If instructed by the RSO, the user should inspect the source/device for damage (otherwise, the RSO may need to travel to the site to inspect the source/device). The user must maintain constant surveillance of the area of the emergency.

- If authorized by the RSO and if the damage to the source/device/container or contents appears to be minor (and, where applicable, the source is secured in its safety position), the device should be placed in its transport container and returned to the storage area for an evaluation and emergency leak test.

- If the source/device/container appears to be severely damaged, i.e. broken, crushed or burned, a barrier at a greater distance around the source should be established (using appropriate warning resources) and all persons should be barred from the area. The user or licensee's personnel must remain physically present at all times to prohibit entry by unmonitored/unauthorized persons.

Management (RSO)Responsibility

- The RSO or his/her designee should contact the agency, police, fire department, manufacturer representative, and consultant, as necessary or appropriate, and report the circumstances of the emergency.

- The company must take action to safely withdraw the damaged camera/source/device and remediate the site upon consultation with the manufacturer's representative, the agency, and consultant.

- The RSO must provide the required written report to the agency (generally a 30 day requirement) describing the circumstances of the emergency, the corrective steps taken, and remediation performed. The RSO should also provide any follow-up reports. Agency requirements for reporting should be followed.

*Note: **Agency** means the federal or state agency (usually radiation control) with jurisdiction and **RSO** means radiation safety officer.*

B. Emergency Due to Radiation Device/Equipment Malfunction

- (Examples might be emergencies due to a disconnected radiographic source, a damaged source or source container, portable gauge with source not retractable, etc.)

User Responsibility

- The user should immediately survey the area with an operable survey instrument and establish a boundary around the area using a predetermined radiation level limit.

- The user should place "Caution—Radiation Area" signs at the boundary, if available.

- The user in charge should maintain control of the area and should not, under any circumstances, leave the area unattended.

- The user should maintain constant visual surveillance of the area in order to prohibit entry by any unauthorized persons.

- The user should send another worker to contact the RSO, or, if not possible, send a person with a written message to make the contact.

- The user should contact the company's consultant and/or the appropriate or radiation regulatory agency, in the event the RSO cannot be readily and timely contacted.

Management (RSO) Responsibility

- The RSO must provide oversight and guidance.
- The RSO must determine if corrective actions are necessary—implementing them if so.
- The RSO must provide the appropriate written report(s) to the Agency within 30 days.

C. Emergency Procedure for Theft of a Source or a Device Containing Radioactive Material

User Responsibility/Procedures

- The RSO, agency, and police must be immediately notified.
- The user should act upon instruction and guidance from the RSO.

Management (RSO) Responsibility

- The company must take the steps necessary to warn the public and recover the camera/source/device, such as providing news releases, media advertisements, etc.
- The RSO should provide the appropriate written report(s) to the agency (usually within 30 days).

D. Emergency Procedure for Loss of a Source or a Device Containing Radioactive Material

User Responsibility

- The RSO, agency, and police must be immediately notified.

- The user should act upon instruction and guidance from the RSO.

Management (RSO) Responsibility

- The RSO or his/her designee should retrace all activities leading to the loss and make every attempt to locate and recover the source.
- If the source/device is not found within 24 hours of the determination of the loss, the company must take the steps necessary to warn the public and recover the camera/source/device, such as providing news releases, media advertisements, etc.

E. Emergency Due to X-ray Machine Equipment Malfunction

User Responsibility

- Power to the unit should be immediately disconnected, the control panel locked, and the key removed.
- A warning sign or label should be placed on the control panel to indicate the unit cannot be operated.
- The RSO should be immediately notified.
- Prevent machine operation until the necessary repairs are performed and the unit has been inspected and determined to be in proper and safe operating condition.
- Maintain direct surveillance of the area to prevent tampering until assistance arrives.

Management (RSO) Responsibility

- The RSO must provide oversight and guidance (even if not present at the accident location).
- The RSO must determine if corrective actions are necessary—implementing them if so.

- The RSO must provide the appropriate written report(s) to the agency (usually within 30 days).

F. Emergency Due to Vehicle Accident

User Responsibility

- The user/driver should determine the condition of the source and container(s) by performing a radiation survey and visual inspection of the device.

- For sources found to be in a shielded position, the user/driver should remain with the exposure device and, if local emergency personnel are not available, send a person to notify the RSO.

- For sources determined NOT to be in a shielded/safe position, or if the survey instrument is not operational, then unnecessary/unauthorized persons should be moved as far from the vehicle as possible (say 100 or more feet—depending on the type of source), direct surveillance should be maintained over the area to prohibit entry, and, where possible, a warning barrier should be established with tapes, ropes, signs, etc. If local emergency personnel are not available, a person should be sent to notify the RSO.

- For any accident, the user/driver in charge should, UNDER NO CIRCUMSTANCES, leave the area/vehicle/radiation source unattended.

- For a minor accident when no radiation hazard exists (no damage to radioactive contents or containers) and the vehicle is movable, no restriction of the area is necessary and the RSO should be notified of the accident as soon as possible.

- The identity of any person present at the site of the accident that may have been exposed to radiation should be documented by the user/driver and reported to the RSO to allow a determination whether (excessive) exposures may have occurred.

Management (RSO) Responsibility

- The RSO should provide oversight and guidance.

- The RSO should determine if corrective actions are necessary—implementing them if so.
- The RSO should evaluate the possible exposure of other persons at the accident site.
- The RSO should provide the appropriate written report(s) to the agency (usually within 30 days).

G. Emergency Causing Exposure of Unmonitored Person (non-radiation worker)

Note: This often occurs when a person unknowingly "wanders" into an area where radiation is in use.

User Responsibility

- Upon discovery or suspicion of exposure of an unmonitored person(s), the source should be immediately returned to the shielded position, or, if using an x-ray machine, the power source to the unit disconnected and the control panel locked.
- If the source cannot be returned to the shielded position, or if radiation cannot be terminated by turning the x-ray unit off, use the previously discussed procedures for the particular condition.
- Obtain and document the names and addresses of all un-monitored persons that MAY have been exposed.
- Notify the RSO and proceed with his/her instructions—if not available, notify the company's consultant.

Management (RSO) Responsibility

- The RSO should provide oversight and guidance.
- The RSO should determine if corrective actions are necessary—implementing them if so.

- The RSO should evaluate the possible doses of persons potentially exposed.

- The RSO should provide the appropriate written report(s) to the Agency within 30 days.

H. Emergency Involving Fire

User Responsibility

- The source should be immediately returned to the shielded position or, if using an x-ray machine, power to the unit disconnected and the control panel locked.

- If the source cannot be returned to the shielded position or power to the x-ray unit cannot be disconnected, fire fighting personnel should be notified upon their arrival of the condition of the radiation source and of areas of potential danger.

- After the emergency, the radiation source/device should not be used until fully inspected according to the company procedures and found to be in proper and safe working condition and after a leak test shows that the radiation source is intact.

Management (RSO) Responsibility

- The RSO should provide oversight and guidance.

- The RSO should determine if corrective actions are necessary—implementing them if so.

- The RSO should provide the appropriate written report(s) to the Agency within 30 days.

References

(The following were available on Amazon.com at the time of publishing)

Atoms, Radiation, and Radiation Protection, 3rd edition. James. E. Turner, © 2007, Wiley VCH. ISBN 978-3-527-40606-7

Design and Evaluation of Physical Protection Systems, 2nd edition, Mary Lynn Garcia, © 2008, Butterworth-Heinemann. ISBN 13:978-0-7506-8352-4; ISBN 10: 0-7506-8352-X

Environmental Radioactivity from Natural, Industrial, & Military Sources, Merril Eisenbud, Thomas F. Gesell, © 1997, Morgan Kaufmann Publishers. ISBN 0122351541

Introduction to Health Physics. H. Herbert Cember, © 2001, McGraw-Hill Health Professions Division. ISBN 0071054618

Radiation Protection: A Guide for Scientists, Regulators and Physicians, 4th edition. Jacob Shapiro: © 2002, President and Fellows of Harvard College. ISBN 0-674-00740-9

Index

D

E

F

S

T

U

V

W

X